Metallurgy of Lightweighting

David J. Fisher

Published by **Materials Research Forum LLC**
Millersville, PA 17551, USA

Published as part of the book series
Materials Research Foundations
Volume 133 (2022)
ISSN 2471-8890 (Print)
ISSN 2471-8904 (Online)

Print ISBN 978-1-64490-212-7
ePDF ISBN 978-1-64490-213-4

This book contains information obtained from authentic and highly regarded sources. Reasonable efforts have been made to publish reliable data and information, but the authors and publisher cannot assume responsibility for the validity of all materials or the consequences of their use. The authors and publishers have attempted to trace the copyright holders of all material reproduced in this publication and apologize to copyright holders if permission to publish in this form has not been obtained. If any copyright material has not been acknowledged, please write and let us know so we may rectify in any future reprint.

Distributed worldwide by

Materials Research Forum LLC
105 Springdale Lane
Millersville, PA 17551
USA
http://www.mrforum.com

Printed in the United States of America
10 9 8 7 6 5 4 3 2 1

Table of Contents

Introduction

The existential threat of global warming has already triggered an urgent movement away from fossil-fuel powered and towards electrically-powered vehicles; it being assumed always that the electricity can be sourced from non-fossil origins. It is observed[1] that, over time, the fuel consumption of various electric-powered vehicles is expected to decrease by 40 to 50%, with a reduction of 45 to 55% in manufacturing costs. Electricity usage arising from light-duty vehicle transport is expected[2] to be between 570 and 1140TWh in 2050; 13 to 26% of the total electricity demand. This increase in demand could correspond to a decrease in light-duty vehicle greenhouse-gas emissions of up to 80%. An already outdated study[3] of energy consumption and greenhouse-gas emissions by a passenger car in China yielded figures of about 91.1GJ and 11.5ton per vehicle, respectively. In the case of China's fuel-consumption rate estimations however the passenger-vehicles curb-weight based consumption-rate targets are specified in a stepped pattern. This can have a marked impact up the manufacturers' light-weighting strategy. This impact was quantitatively evaluated on the basis of China's domestic automotive market. From the cost-effectiveness point-of-view, it could be demonstrated – with stepped fuel consumption rate targets – manufacturers have a strong motive for manipulating curb-weights so as to qualify for more favorable targets. In the case of China's 2010-2014 domestic vehicles there was a markedly imbalanced curb-weight distribution, with a considerable number of vehicle models clustered at the target-preferred end of each weight class. The models which were most likely to have been manipulated were identified[4], and accounted for some 10% of the total. It was concluded that, with a shift from stepped targets to smooth targets, the affected models would have an average 17.92kg of mass reduction and a 0.073l/100km fuel consumption-rate improvement. The stepped targets may have deterred manufacturers from implementing lightweighting changes.

Transportation is responsible for over 30% of US greenhouse-gas emissions, and improvements in vehicle design must be aimed at improving fuel (table 1) economy by using advanced power-trains, lowering rolling-resistance and decreasing aerodynamic drag.

The problem can also be attacked from a different angle however by simultaneously reducing the weight of the vehicles; a process known as lightweighting. The lower weight obviously reduces the amount of energy which is consumed and thus reinforces the beneficial effect of the move away from fossil fuels. It has been estimated that every 50kg of weight-deduction from that of the average 1500kg vehicle reduces its CO_2 emissions by between 4 and 5g. An emphasis on ultra-lightweight and ultra-low drag can reduce the required propulsive power by some two-thirds[5]. This can make direct-hydrogen fuel-cells and hydrogen gas-storage more practical. Other strategies, beside lightweighting via material substitution, include[6] increased recycling, vehicle down-sizing and vehicle-sharing. Down-sizing and more intensive use have the greatest potential mitigation effect, but are rather idealistic and may encounter considerable public resistance. The adoption of all possible mitigation strategies could reduce emissions by up to 57% over the life-cycle of a single vehicle.

An obvious manner in which to reduce the weight is to employ less-dense materials, such as polymers, carbon fibers and composites; although this brings its own complications. The present work concentrates however on the use of metals; either low-density alloys or higher-density alloys whose superior strength permits a reduction to be made in the volume of metal which is required.

Another aspect not considered in depth here is the somewhat peripheral subject of tool-lightweighting. That is, whatever metallurgical changes are made in order to minimise vehicle weight, most of the components still have to be machined at some stage. Machine-tools, such as shaping machines, can be very heavy. Secondary savings can therefore be made by lightweighting even the machine tools which are used to achieve lightweighting. Much of the environmental footprint of the tool is due to the energy which it consumes during use, and this is related to the weight of the work-table and the work-piece, plus the cutting force. Study of a 3-axis vertical milling machine[7] has shown that savings of up to 38% can be achieved with regard to the energy required to move the work-table.

The lightweighting option was foreseen[8] some decades ago, when enthusiasts of electric vehicles realized that the wide use of plastics could offset the weight of the batteries of those days, which then accounted for some 30% of the total weight. Since those days, battery power has improved enough to make electricity a truly alternative fuel, and the current price is of the order of $137/kWh. An early study noted that the use of plastics led to a parts-reduction ratio of 5:1, that tooling costs were 60% lower than for steel dies and that the use of adhesives cost 25 to 40% less than did welding. New grades of ultra high-strength steels can be made thinner and thus save 30% in weight by replacing conventional mild steels, although more expensive processing may be required. Engine-block structures are now made from aluminium rather than grey cast iron, but compacted-graphite cast iron has a higher strength at high temperatures and pressures so that engines can be made smaller and lighter than aluminium engines. There is a correlation between the mechanical properties of a steel and the resultant weight of vehicle components. New forging materials and forging technologies can greatly contribute to lightweighting. The use of high-quality steel in transmission applications can be particularly beneficial.

Table 1. Simulated fuel-use predictions for various lightweighting strategies

Vehicle Type	Strategy	Fuel Economy (mpg)
compact	*status quo*	33.0
compact	6% high-strength steel	34.2
compact	6% aluminium	34.2
compact	11% aluminium	35.2
compact	19% high-strength steel	36.7
compact	23% aluminium	37.5
mid-range	*status quo*	30.5
mid-range	6% high-strength steel	31.5
mid-range	6% aluminium	31.5
mid-range	11% aluminium	32.5
mid-range	19% high-strength steel	33.9
mid-range	23% aluminium	34.7
luxury	*status quo*	21.5
luxury	6% high-strength steel	22.3
luxury	6% aluminium	22.3
luxury	11% aluminium	22.9
luxury	19% high-strength steel	23.9
luxury	23% aluminium	24.5

The use of use of high-strength steel and advanced high-strength steel in automotive vehicle structures has proved to be one of the most feasible means for lightweighting vehicles. Automotive body structures made from high-strength steels can have a reduced thickness and thus make possible the development of light-weight fuel-efficient vehicles. Changing the body structure can also reduce the body-structure weight but such changes lengthen the product-development time, require re-tooling and increases costs. Tailor-made parts, in whereby pieces of various thickness are combined into a single, involve no design-changes and fewer tooling requirements, and are a good alternative to reducing greatly the weight of the body structure. Lightweighting optimization studies have shown such alternatives are proven techniques. It is important to identify potential areas for potential tailored-part incorporation, with such parts including linear/curvilinear tailor-welded blanks and tailor rolled blanks. The introduction of ultra high-strength steels can however result in the significantly increased wear of stamping tools; something which is difficult to predict can lead to unexpected process failures. It is therefore important to monitor the onset of severe wear. The aim of active monitoring is to detect the initial onset of a change of state by measuring variables such as punch-force and acoustic signals.

Reductions in sheet thickness due to the use of high-strength steel increasingly exploited as an effective means for weight reduction because of the cost-savings. It can be shown that steel is the only material which is suitable for satisfying the conflicting demands of required properties and weight-saving. A typical example of component substitution is the use of S650MC high-strength steel for long-member applications in commercial vehicles. On the other hand the ability to introduce new lightweight materials into vehicles is not entirely simple, it is also necessary to consider the feasibility of manufacture. Improved manufacturing technologies are often required when working with high-strength steels, particularly because of their reduced formability and weldability. Laser-based material processing, such as laser-welding offers many possibilities[9]. Hot-stamping and roll-forming are additional possibilities for advanced manufacture.

A simulation approach has been proposed[10] for the formation of hollow gears from EN-10130 steel by using a tube hydroforming process. Manufacture of the gear teeth was analyzed using forming-limit diagrams and a strain non-uniformity index method. The pressure and compression-rate were controlled independently in order to produce defect-free products. The forming-limit diagram and strain non-uniformity index method showed good agreement in predicting the failure of the part. Ideal parameters could be proposed to get a defect-free part. The resultant gear could be for low torque transmission application. Tube hydroforming is a technology which can efficiently reduce the weight of a component or assembly while increasing the strength. The process produces parts having a high degree of complexity, such as differing cross-sections in a single piece. Tube hydroforming involves the expansion and sizing of tubes within a closed die under action of a pressurized fluid, with simultaneous axial or radial compression. The best results are obtained by using an optimum combination of internal pressure, feed, friction, load and blank thickness. Hydroforming has been used to reduce the mass of chassis cross-members, crash members and exhaust housings.

There have been attempts to replace existing gear-shift forks, currently made from ferro cast iron, with aluminium die-castings without compromising strength and stiffness. The die-cast aluminium had good mechanical and thermal properties when compared with those of the ferro cast iron. A six-speed manual transmission was also used as an example of the simulation and validation of new designs. Advanced linear topology optimization methods were suggested to be the most promising techniques for light-weighting design of power-train structures.

Life-cycle assessment[11] involves considering the entire vehicular life: all the way from mineral extraction, through component manufacturing, use of the vehicle and disposal or possible recycling. With regard to manufacture, aluminium components have the greatest impact because of the high energy expenditure which is required for aluminium extraction. This stage has little effect upon the overall life-cycle analysis, because the benefits of weight-reduction during use of the vehicle make a much greater contribution. Urban driving has the most relevant impact, because of frequent starts and stops and long periods of idling. Non-urban driving is less demanding, because of higher average speeds and fewer halts.

A thorough assessment of life-cycle effects involved in vehicle lightweighting (figure 1) requires a rigorous evaluation to be made of weight-induced consumption, upon which any energy and sustainability benefits enjoyed during use stage directly depend. An estimate has been made of the weight-related energy consumption of pure electric vehicles. The correlation between consumption and weight is expressed by the energy reduction value coefficient, which quantifies the specific consumption-saving which can be achieved by a 100kg weight-reduction. The energy reduction value has been estimated for a number of electric vehicle models and 3 driving behaviors. The energy reduction values were found to range from 0.47 to 1.17kWh/(100km100kg), with the variations depending mainly upon the vehicle size and driving behavior. The analytical method was refined so as to estimate accurately the energy reduction value for any real-world electric vehicle, given the vehicle technical specifications. An evaluation was also made of the environmental implications of lightweighting by defining an impact reduction value. This was estimated for 3 distinct electricity grid mixes.

☐ Hot-formed steel sheet

☐ Cold-formed steel sheet

☐ Aluminium sheet

☐ Extruded aluminium profile

☐ Cast aluminium

Figure 1. An example of lightweighting material choices in the design of the Porsche 800V Taycan electric sports car. Reproduced[12] under Creative Commons Attribution Licence (with modified key)

Electrification and lightweighting are important in reducing greenhouse-gas emissions from light-duty vehicles, and it is essential to make cradle-to-grave life-cycle (table 2) assessments. A regional life-cycle assessment of lightweighting and electrification can reveal differences in local climates and driving patterns, and lightweighting can accentuate such regional differences. In the case of mid-sized vehicles, aluminium-lightweighted vehicles with internal combustion engines are expected[13] to have similar emissions to those of hybrid electric vehicles in some 25% of US counties, and to be lower than those of battery electric vehicles in 20% of US counties. As well as assessing the life-cycle details of location, it can be useful to assess the life-cycles of individual components (tables 3 and 4).

Table 2. Life-cycle greenhouse-gas emissions for various vehicles

Vehicle	Emissions ($kg_{CO_2}[eq]$)
diesel	32630-50261
hybrid electric	30179-46560
plug-in hybrid electric	25859-40573

In an early example, a lightweight frameless steel door design achieved a 42% weight-saving over the average frameless door and a 22% weight-saving over the very lightest framed door. The appreciable weight-savings were made by using high-strength and ultra-high strength steels, combined with the use of tailored blanks and hydroforming. The door outer panel was first made of stamped 0.7mm bake-hardenable 260 sheet steel. During the development of the frameless door, a further weight-reduction was achieved by using sheet-hydroforming for the door outer panel. Door structures could later be successfully manufactured by using 0.6mm dual-phase 600 hydroformed steel outer panels, leading to additional weight-savings. In recent years however, it has been decided that it is not acceptable to obtain such weight-savings without considering the effect, of the means used, upon the environment as a whole.

Table 3. Cost-benefit analysis for manufacturers

Lightweighting Strategy	Extra Cost ($)	Greenhouse-Gas Reduction ($/tonne$_{CO_2}$[eq])
6% high-speed steel	400-1031	148-534
6% aluminium	406-621	148-277
11% aluminium	640-1241	128-287
19% high-speed steel	2752-3360	352-579
23% aluminium	2844-3516	315-431

For example, the development[14] of a new ultralight-door, that led to a 40% overall mass reduction, was subjected to life-cycle assessment. Ultra-light doors are found to be $6.0g_{CO_2}[eq]/km$, and 86kJ/km better, and these figures can be further improved, by powertrain-resizing, to 2.8g$_{CO2}$[eq]/km and 40kJ/km, respectively. The use of an aluminium-intensive door design leads to a weight-reduction of some 49.5kg for a 4-door vehicle. The lightweighting possibilities offered by advanced high-strength steels for exterior panels have been studied[15] with particular regard to dent-resistance, due to this being of great everyday importance. Alloys which exhibit a good surface quality and high strength were compared. Denting tests were performed on plates having various curvatures which simulated damage to door, roof and hood panels. The application or 1 or 2% of pre-strain and differing baking treatments were also considered. The minimum gauge which was required for optimum dent resistance was thus obtained. Hierarchical analysis was applied[16] to the design of a door-shaped structure. Various boundary conditions and loading ranges were established in order to solve analytical problems of natural frequencies and stress of the transverse beam. These were combined with a genetic algorithm in order to perform lightweighting calculations and obtain the optimum design.

An estimate was made[17] of the lightweighting possibilities for the frame structure of a commercial road vehicle. This was based upon simplified static-load cases which play a predominant role in the dimensioning of a frame structure; thus avoiding calling the overall validity of the conclusions into question. A comparison of various materials showed that light metals did not offer any weight-reduction advantage over steel. A material-independent topology-optimization approach offered a superior weight-reduction potential than did a simple change of material. Because of constraints of complexity, directly linked to production and assembly costs, the ladder frame=structure is the current state-of-the-art design. Practical examples showed that the weight of a commercial vehicle could be reduced by 10% and that the main parts of the frame structure could be reduced by 30% by using high-strength steel and roll-forming.

An often-overlooked component for potential lightweighting is the starter motor, and any attempt at lightweighting must consider aluminium. The starter motor is fitted directly onto the engine, and an aluminium housing provides structural stability. It also aids heat-dissipation because of its good thermal conductivity as well as providing a grounding path for electrical current. Aluminium constitutes 20 to 25% of the starter motor weight. No significant weight-reduction is possible unless the die-cast housing and end-plate are considered. The possible alternatives include plastics, magnesium alloy and composites, but account must taken of their structural stability, strength, and electrical and thermal conductivities.

A particular study was made[18] of minimizing the weight of a hood assembly while meeting strength and stiffness targets. The design variables which were considered were the thickness of the panels, and design constraints were set for stresses and stiffness on the basis of a design verification plan. Data response surfaces were generated by using various algorithms for predicting structural performance parameters such as displacement, modal frequency and stress. A weight-saving of 1kg was finally estimated.

Even the wiring harness of a vehicle can be subjected[19] to life-cycle assessment, after replacing the usual copper wire with a copper-tin alloy. Such a cable reduces the weight and volume by 53% and 41%, respectively. The resultant environmental burden could also be 54% lower, with the exhaust emission reduction being 160kg$_{CO_2}$[eq]. Reduction of the wire diameter could also make the cabling more sustainable from the environmental point-of-view.

Table 4. Cost-benefit analysis for consumers

Lightweighting Strategy	Fuel Saving (gal)	Cost Saving ($)
6% high-speed steel	215	765
6% aluminium	215	605
11% aluminium	626	2240
19% high-speed steel	1009	3657
23% aluminium	1198	4344

A study has been made[20] of vehicle lightweighting with regard to the impact of end-of-life recycling. Changes in material composition were assessed by including the entire life of the vehicle: production, with its associated energy costs, the use-phase with its weight-related energy consumption and end-of-life recycling with parts re-use, material recycling and perhaps energy-recovery. The use of aluminium, advanced high-strength steels and carbon-fiber reinforced plastic was compared. The results showed that the effects of vehicle-lightweighting upon the production and end-of-life phases were as important to consider as the benefits enjoyed during the use-phase. Material-lightweighting had to be analyzed together with possible recycling because a maximum life-cycle energy-reduction of 23.3MJ/kg of lightweighting was expected. A suitable combination of the above phases could increase the benefits to 51.0MJ, while an ill-judged combination could lead to an energy-consumption increase of 92.7MJ. An overall reduction in vehicle impact of between 7 and 14% was predicted[21] by combining all of the best alternatives for each component. Chassis improvement alone provided 51% of the total reduction. Structure-lightening by material substitution was found to be fundamental in reducing fuel and energy consumption during the use-phase (tables 5 and 6). It was significant

Materials Research Forum LLC
https://doi.org/10.21741/9781644902134

however that the modest lightening effect which is provided by aluminium is in fact better than the larger one that is provided by carbon-fiber and magnesium use, where the greater impact resulting from their extraction and end-of-life treatment has a negative effect. High recycling-rates of energy-intensive lightweight materials are important in order to benefit fully from vehicle lightweighting, but the quality of recycled scrap has be kept as close as possible to that required by the original specification. For conventional vehicles, the use-stage accounts for some **85%** of the total life-cycle energy consumption and greenhouse-gas emissions[22]. Power-train design and efficiency, as well as fuel-type, largely control this. Some **85%** of the materials are theoretically recyclable, but the recovery of plastics and the separation of alloy components require further research.

Table 5. Properties of common lightweighting alloys

Alloy	E (GPa)	YS (MPa)	UTS (MPa)	e (%)	Density (g/mm^3)
DP1000	200	669	985	10.25	7.82
Fortiform 1050	190	772	1249	10.57	7.78
Usibor 1500	171	1100	1546	3.95	7.76
AA7075-T6	56	531	670	13.83	2.81
AA6111-T6	78	381	485	10.92	2.67
AA6111-T4	51	144	327	18.56	2.67

Use of the solution heat-treatment, forming and in-die quenching process permits the construction of large sheet metal parts without any need for rivets or other bonding means that could complicate end-of-life disassembly and recycling[23]. A life-cycle assessment showed that the added energy which is required for the above process is more than compensated by the resultant benefits of enhanced end-of-life recyclability.

Table 6. Properties of common non-ferrous lightweighting alloys

Alloy	0.2%YS (MPa)	UTS (MPa)	Specific Strength (kNm/kg)
AA6063	195	258	76
AZ31	203	247	121
AM-EX1	184	259	116

Use-phase fuel consumption accounts for most of a vehicle's life-cycle energy consumption and greenhouse-gas emissions. The use of thin-wall ductile cast iron is a lightweighting option which can provide a comparable weight-reduction to that offered by aluminium while offering superior mechanical properties. A parametric life-cycle model has been used[24] to assess the life-cycle performance of thin-wall ductile cast iron in comparison with conventional cast iron and cast aluminium as regards energy and carbon dioxide production. The model was applied to the replacement of a differential casing, an engine block and other cast-iron parts. Lightweighting by 2% resulted in equal life-cycle energy and greenhouse gases for thin-wall ductile cast iron and conventional cast iron, while 37% of lightweighting was required in the case of thin-wall ductile cast iron in order to equal the cast-aluminium performance. The assembly of divers lightweight materials may also require new joining techniques. The incidence of corrosion also has to be considered, simply because of the close proximity of very different materials.

Table 7. Cost benefit of material replacement

Replacement Strategy	Weight-Reduction/Cost-per-Part
alloy-steel by titanium	4.5-37
steel by graphite-fiber reinforced plastic	5-30
mild steel by high-strength steel	10
carbon-steel by stainless-steel	12-38
aluminium by magnesium	17-35
steel by glass-fiber reinforced plastic	17-35
steel by aluminium-matrix composite	17-43
cast iron by aluminium-matrix composite	17-43
mild steel by aluminium	20-46
cast iron by aluminium	20-46
steel by magnesium	24-50
cast iron by magnesium	24-50

It is possible to reduce weight by about 25kg by integrating an aluminium cylinder-head and intake manifold for light-duty commercial vehicles, leading to savings of about 2g of

CO_2. The existing cast iron was first analyzed for various worst engine loading conditions in order to determine the stiffness at various known locations. The aim[25] was to have a new design which had the same stiffnesses, in the same locations, as those which were present in the cast-iron version. By using the integrated aluminium cylinder head and intake manifold it was estimated that cost savings of up to 30%, a reduction in weight of 40% and an improvement in fatigue performance of 25 to 30% could be achieved. Because the piston-assembly has a direct influence upon the friction losses in an engine, attempts have been made to reduce the reciprocating mass and to reduce friction. Improvements in design and the use of new surface coatings for pistons and rings have significantly reduced friction and improved fuel economy[26].

A life-cycle approach was used[27] to estimate the energy-based, environmental and other benefits. The costs (table 7) included estimates of the expense required to commercialize new technologies. When benefit/cost ratios were calculated, on the basis of various assumptions, the ratio was always appreciable. It was also expected that lightweighting might encourage the development of hydrogen fuel-cell powered vehicles. The use of hydrogen raises other problems because of its safety risks. A study was made[28] of the safety of a lightweight hydrogen fuel-cell propelled vehicle with an overall unloaded weight of less than 800kg and a range of 180 miles. It contained a hydrogen fuel tank at a pressure of 350bar. It was impacted head-on at velocities of up to 40km/h and it was concluded that the crashworthiness was good.

A weighted average is generally used to calculate the impact of the production of an alloy for a given input-mass. The latter is related to the masses of the individual materials in their finished form following melting, casting and machining. The impact per kg of product for these processes is calculated in terms of the fuel-reduction value. The fuel consumption for a conventional vehicle of a given mass is calculated, and the fuel-reduction value then specifies the mass-dependent fuel consumption. The vehicle-part share of the mass-dependent component of the fuel consumption is then a function of the part-mass. The change in fuel consumption of the vehicle, and the change in fuel consumption of the part due to lightweighting, is calculated and the fuel consumption for the lightweighted part, relative to its original value, is found. Following this somewhat laboured reasoning, the fuel-consumption for the lightweighted vehicle is equal to its original consumption minus the change in fuel-consumption.

The initial lightweighting step frequently leads to a concomitant re-sizing of the power-train, thus further reducing the weight. In the absence of power-train re-sizing, a typical fuel-reduction value might be of the order of $0.2l/(100km 100kg)$. The total energy-use and greenhouse-gas emissions are based upon both the up-stream (extraction, refining, delivery) aspects of fuel-use and its in-use combustion.

It is found that scrap motor vehicles are 85% recyclable, but that such recycling will involve transport (over perhaps 100 miles), dismantling and shredding; the efficiency of the latter process being of the order of 95%. Most of the aluminium (up to 70%) which is present in automobiles is recycled, and it can be assumed that the amount of secondary aluminium in new parts is of the order of 65%. Assuming that the production-energy is 126MJ/kg for new aluminium and 18MJ/kg for secondary aluminium, the greenhouse-gas emissions resulting from production are of the order of 8kg$_{CO2}$[eq]/kg for primary aluminium and 1.1kg$_{CO2}$[eq]/kg for secondary aluminium; casting and machining energies and greenhouse-gas emissions being essentially the same for both materials. So finished cast aluminium products made from new aluminium have an energy-impact of the order of 150MJ/kg, with associated greenhouse-gas emissions of the order of 9kg$_{CO2}$[eq]/kg. The equivalent values for finished cast products made from recycled aluminium are some 31MJ/kg and 1.9kg$_{CO2}$[eq]/kg respectively.

A 10% weight-reduction in all of the cast-iron components of a vehicle is equivalent to a 1.03% weight-reduction of the entire vehicle. The material-production, plus manufacturing, energy which is associated with conventional cast iron is about 3980MJ. When replaced with thin-wall ductile cast-iron, the production-energy rises to 4310MJ. Upon replacement by cast aluminium, the energy is 5650MJ. The higher energy which is involved in the case of the thin-wall iron is due to the alloying-element content.

The amount of fuel which is consumed is 30600MJ for conventional cast-iron, and 27500MJ for the thin-wall cast-iron; the 10% reduction being due to a reduction in fuel consumption from 762 to 686*l* of petroleum. Replacement with cast aluminium further reduces the fuel consumption to 419*l* and the in-use energy to 16800MJ; a reduction of 45% when compared with conventional cast-iron use. The reductions of in-use energy, for both thin-wall cast-iron and cast aluminium are sufficiently large to justify the increased production energies for these alloys. The reduction in total vehicle fuel consumption is 0.3% when all of the cast-iron components of the vehicle are replaced with thin-wall cast-iron, and is 1.5% in the case of replacement with cast aluminium.

The reduction in energy which occurs during the use-phase when employing lighter alloys is sufficient to ensure a net gain in total life-cycle energy. The thin-wall cast-iron has an associated total life-cycle energy use of 31900MJ, as compared with 34600MJ for conventional cast-iron. Cast aluminium has an associated total life-cycle energy use of 22500MJ. The energy reduction which results from the replacement of cast-iron components is only modest however because cast-iron accounts for only some 10% of the total mass of the vehicle.

Consideration has also been given[29] to life-cycle emissions, assuming the use of both aluminium and high-strength steel and degrees of lightweighting of between 6% and 23%. By including the emissions which were associated with the production of aluminium and high-strength steel, a comparison could be made between the increased emissions that were associated with vehicle production and the reduced emissions which resulted from lightweighting. It was estimated that 2 to 12 years of greenhouse-gas offset would result from 6% of lightweighting while 4 to 9 years of greenhouse-gas offset would result from 23% of lightweighting. The cost was also lowest in the latter case. It was noted[30] that although weight-reduction could incur additional expense, this could be offset by secondary weight-savings which were made in sub-systems such as the power-train[31]. Popular metallic materials for lightweighting include aluminium and magnesium, while the increasing interest in hybrid vehicles increases the need for lithium, cobalt and nickel. Replacing recyclable steel with the lighter materials reduces the recycling-rate of vehicles unless the lightweighting materials are also recycled. Recycling of those materials thus becomes essential to sustainable growth.

High-pressure die-casting of aluminium alloys is one of the most common methods used for the high-volume production of lightweighted components and permits high production-rates and very near net-shape component manufacture. Its use ensures weight-reductions of between 30 and 50% when replacing conventional steel components.

The amenity of aluminium alloys to high-pressure die-casting is relevant to the manufacture of components such as sub-frames, beams, drive-trains, clutch housings and battery housings. The sub-frame carries the engine, drive-train and suspension. Typical reductions are expected to be of the order of 40% in thickness and 20% in mass. The use of vacuum-casting has advantages if the numbers of components to be produced is sufficiently high; the investment in the required new equipment will otherwise make the process uneconomical. It is suggested that it becomes economical when some 8000 components per annum are to be produced, and super-vacuum die-casting then offers a lower-mass component at some 6% lower cost than that of high-pressure die-casting. The overall reduction in cost for an automobile sub-frame component can be about 5%. The material costs are more significant for larger components, and the savings which result from super-vacuum die-casting are also more significant in the case of larger parts. At a rate of production of 100000 components per annum, the reductions in mass might typically be 15% for small parts, 12.5% for medium-sized parts and 12.5% for large parts. At a rate of production of only 5000 components per annum, the latter figures may typically become 45%, 35% and 30%.

A life-cycle assessment was made[32] of the particular case of wireless-charged electric vehicles in comparison with plug-in vehicles. The results showed that a wireless-charged battery could be downsized to between 27 and 44% of a plug-in battery. An associated reduction of 12 to 16% in vehicle weight in the wireless case could effect a reduction of 5.4 to 7.0% in battery-to-wheel energy consumption. The wireless system consumed at least 0.3% less energy and emitted 0.5% less greenhouse gas than did the plug-in system over the total life-cycle. If the wireless-charging efficiency could be improved to the same level as the assumed plug-in charging efficiency, the difference in life-cycle greenhouse-gas emissions for the two systems was expected to increase to 6.3%.

Table 8. Energy cost of US steel production

Sub-Process	Energy (MJ/ton)
iron-ore extraction	2090
coke production	18110
sintering	70
blast furnace	2070
basic oxygen furnace	804
hot-rolling	1550
cold-rolling	1630
galvanizing	810
stamping	1000

Even in the case of non-electric vehicles, lightweighting directly reduces petroleum-dependence and its associated strategic concerns by decreasing engine, rolling-resistance and braking losses as well as indirectly doing so by permitting a smaller but more efficient engine to maintain the same performance. A comparison of conventional engines, hybrid electric vehicles, plug-in hybrid electric vehicles and battery electric vehicles indicates that weight-reduction has its greatest effect upon the efficiency of conventional vehicles. On the other hand, lightweighting of the other types of vehicle had a greater effect in reducing component and overall costs (tables 8 and 9). Lightweighting was once expected to be cost-effective when the expense was less than $6 per kg of reduced weight. In the case of battery-powered electric vehicles, the latest improvements in battery energy-density and electric motor capabilities have progressed to the point where expensive lightweighting materials are less needed. A related comparison was made[33] of an advanced high-strength steel and aluminium. Under current economic conditions, the steel version costs $595 less per vehicle than did the aluminium version. The cost-advantage of the steel lay in the vehicle body while the cost-advantage of the aluminium lay in the battery, motor and chassis. With the lightweighting advantages of the battery and motor declining, the cost-gap was expected to attain a $743 per vehicle advantage for steel.

Table 9. Energy cost of US aluminium production

Sub-Process	Energy (MJ/ton)
bauxite mining	671
Bayer-process extraction	20700
anode production	1730
Hall-Heroult electrolysis	5440
primary ingot-casting	1180
scrap recycling	1270
secondary ingot-casting	5190

It is to be noted that the terms, 'mass' and 'weight', tend to be used indiscriminately in this field even though, from the physics point-of-view, that might not always be correct. A specifically physics-based method for estimating so-called mass-induced fuel consumption was proposed[34] which used available data to calculate that mass-induced fuel values ranged from 0.2 to 0.5l/(100km100kg). Lightweighting was shown to be most beneficial when applied to vehicles having a high fuel consumption and high power. Later estimates[35] of mass-induced fuel consumption and fuel-reduction values showed that the latter values could be mathematically deduced while avoiding additional testing. The mass-induced fuel and fuel-reduction values which were calculated for 83 vehicles ranged from 0.22 to 0.43 and from 0.15 to 0.26l/(100km100kg), respectively. They increased to between 0.27 and 0.53l/(100km100kg) when power-trains were re-sized in order to ensure equivalent vehicle performances. Use-phase fuel-consumption could be deduced from mass-induced fuel and fuel-reduction values by using life-cycle assessments of vehicle lightweighting. The mass-induced fuel consumption could potentially be deduced from the life-cycle greenhouse-gas emission benefits which accrued from lightweighting a grille using magnesium or carbon-fiber. The physics-based model of mass-induced fuel and fuel-reduction values for internal combustion engine vehicles was extended[36] to electric and hybrid vehicles. The utility of the model was shown by calculating mass-induced fuel and fuel-reduction values for 37 electric vehicles and 13 internal-combustion vehicles. Battery-powered electric vehicles had much smaller mass-induced fuel and fuel-reduction values, ranging from 0.04 to 0.07l/(100km100kg), than those for internal-combustion vehicles, where they ranged from 0.19 to 0.32 and from 0.16 to 0.22l/(100km100kg), respectively. The mass-induced fuel and fuel-reduction values for other electric vehicles tended to lie between those for internal-combustion and battery-powered vehicles. The use of power-train re-sizing increased the fuel-reduction values for non-battery vehicles. The lightweighting of electric vehicles was less effective in reducing greenhouse-gas emissions than was the lightweighting of internal-combustion vehicles, and the benefits differed markedly moreover for different individual vehicles.

Another study[37] compared an internal combustion driven vehicle, a hybrid electric vehicle and a plug-in hybrid electric vehicle. Each vehicle was functionally equivalent and incorporated the structural support which was required for heavier power-trains. The results indicated that the greatest life-cycle energy and greenhouse-gas emission reduction occurred when steel was replaced by aluminium. On the other hand, when the greater energy required to produce aluminium was included, the energy and greenhouse-gas reductions per unit mass were less favorable for aluminium. It was noted that 0.2 to 0.3kg of structural support was required per unit increase in power-train mass. An analytical vehicle simulation, combined with cost-assessment, was used[38] to identify the optimum degree of weight-reduction which was required in order to minimize the manufacturing costs and total costs. The results revealed the existence of significant secondary weight- and cost-saving opportunities in the case of battery electric vehicles, but a higher sensitivity to vehicle energy use for conventional vehicles. Lightweighting could lower vehicle costs, but the outcome was very sensitive to parameters which affected lifetime fuel costs or battery costs. The optimum amount of primary mass reduction which minimized the total cost was similar for both conventional and electric vehicles, and ranged from 22% to 39%; depending upon vehicle range and usage. The difference between the optimum solutions which minimized manufacturing versus total costs was higher for conventional than for battery electric vehicles.

Attention also has to be paid to the bending stiffness of materials, as related to crash-worthiness. The front body-structure of an automobile should have sufficient stiffness and strength to protect occupants during a high-speed head-on collision, but should also not cause undue damage to pedestrians during a low-speed collision. This conflict can be resolved by careful design[39].

Lightweighting efforts aimed at improving fuel economy require the detailed examination of new materials in order to make sure that the efficiency gains do not sacrifice safety. A variable-section and multiple-materials lightweighting design strategy has been based[40] upon collision safety. A racing-car frame was used as an example and the optimum design which provided good collision-safety was identified. The lightest frame-weight was used as the design goal, and optimum latin-square and response-surface methods were used to optimize the thickness of each pipe frame. Assuming that the peak acceleration of the cockpit was reduced by 20.02% and that the degree of intrusion at the brake pedal was reduced by 25.31%, a weight-reduction of 14.38% was to be anticipated. In order to light-weight a truck frame, the stiffness-sensitivity, mode-sensitivity and the compensating sensitivity of the static and dynamic structural responses were analysed[41] by using a size-optimization technology. The total life, crack-initiation life and crack-growth life were also analyzed.

A sandwich-structured vehicle engine-compartment cover was expected[42] to exhibit a uniform stiffness distribution in the transverse direction. Finite-element body-models were used to study the injury mechanisms involved in lower-extremity impact under real collision conditions. It was found that a so-called ringing-effect of the force during impact-testing was due mainly to a system resonance which was amplified by a sharp transition from elastic to plastic yielding in the stress-strain relationship. Various methods and devices could dampen the ringing. By investigating the large-deformation failures of high-strength steels, spot-welds and adhesive bonds, etc., models could be developed for predicting fracture under impact-loading for almost all of the lightweight materials and joint-types used in vehicles. Compression-testing of batteries meanwhile identified the relationships between battery-material failure, internal short-circuiting and temperature rises under mechanical loading. The maximum deflections and strength limits of beam structures having 3 different cross-sections, aimed at weight-reduction, have been studied[43] theoretically for conditions of uniformly-distributed loading with a fixed support at one end and a roller support at the other. This established the lightweighting limit of beam structures having various cross-sections. In the most extreme case, concepts have been investigated[44] for choosing material properties and component configurations in order to mitigate blast loads while nevertheless reducing weight. The dynamic response index was used as an occupant-injury metric for judging the effectiveness of each blast-mitigation configuration. A finite-element model of a V-hull structure was used as an example. The material properties and the configuration of the inner bulkheads which connected a V-shaped outer surface with the inner floor were used as design parameters for reducing the dynamic response index at a typical occupant location. It was found that the weight of the structure and the dynamic response index could both be reduced at the same time by using a structural design that featured energy-absorption and decoupling mechanisms among the bulkheads, floor, seat and occupant.

Tailored welded blanks are an important approach[45] to vehicle-body lightweighting, and their formability is a key factor. Tailored blanks offer considerable lightweighting opportunities and have been applied[46] to the design of an automobile in which all of the front-rail parts were divided into multiple sheets such that the gauge of each sheet was available as a design variable for optimization purposes. The equivalent-static-load method was used for structural optimization, and a standard moderate-overlap frontal collision criterion was used in the non-linear load case. The torsion and bending stiffness of the vehicle body-in-white (plain unfinished chassis state) were used as design constraints. The object was to minimize compartment-intrusion in a moderate overlap frontal collision situation. It was found that the section force on the front rails was reduced by tailored blank design, but the collision-energy which was absorbed was very close to the baseline (over 95%) with lightweighting being achieved. The energy which was absorbed by the entire system was equal to, or greater than, the baseline value. This was attributed to the fact that the tailored front-rails leveraged the energy-absorption of the vehicle as a whole and of the deformable barrier. It was concluded that a 23.59% weight-saving in the front rails could be achieved, by using a tailored design, while improving the collision safety-rating. In similar work[47], a vehicle which met 5-star crashworthiness criteria was developed by using a high-accuracy finite-element model for collisions. By suitable re-design of the front-side rail, sub-frame and front fender, with the acceleration wave-form, front-bulkhead intrusion and the deformation pattern of the front side-rail being used as constraints, the safety of the vehicle could be improved while attaining lightweighting targets. An automotive front fender system consists of a fascia, grill, bumper beam and connectors. The bumper beam is the main structural component which protects the occupants. A static non-linear analysis was made[48] of the front fender bumper beam for various metal choices: steel, aluminium, magnesium and copper, and for various thicknesses: 3 to 6mm, and for velocities of 2.5 to 8kph. The results improved the performance of a vehicle in protect the occupants during low-velocity impacts.

A recent lightweighting design strategy[49] combines experiment and decision tools to improve the efficiency of body-in-white design. A finite-element analysis based upon static-dynamic performance and frontal collision safety of the body-in-white is first carried out, and the correctness of the finite-element model is then confirmed by vehicle frontal collision-testing. The design variables for the front-end structure of the body-in-white are subsequently incorporated by using a contribution analysis technique.

Candidate lightweighting solutions (tables 10 and 11) for the front-end structure of the body-in-white are then obtained via experimental design. A multi-objective decision is then made from among the numerous possible lightweighting solutions in order to obtain the best one. It was concluded that the weight of the front-end structure of the body-in-white could be reduced by 4.43kg, corresponding to a lightweighting of 7.23%, while satisfying baseline requirements. Another step-by-step optimization technique, combined with body-in-white performance-matching was used[50] to find a lightweight-design which fully satisfied all of the static and dynamic requirements of the body-in-white. The results before and after optimization showed that, by making small reductions in the bending and torsional stiffnesses (0.2 and 0.6%, respectively), the length, width and height of the car body increased by 15, 13 and 9mm, respectively, the first-order bending and torsional-mode frequencies rose by 5.6 and 9.2%, respectively, and the weight of the body was decreased by 19.9kg; a lightweighting of 5.76%.

In order to satisfy the requirements of lower weight and crashworthiness, the front-end of a vehicle was optimized[51] by combining material replacement with structural improvements. An aluminium front-end model was compared with the steel equivalent. In order to increase the crashworthiness of an aluminium front-end, multi-cell cross-sections having various cell-numbers were designed, and their energy-absorption capabilities were analyzed under three-point bending and axial crushing conditions. An optimized aluminium multi-cell structure could improve vehicle crashworthiness while markedly reducing the weight of the front end. The structural crashworthiness of a car was optimized[52] under the conditions of a 40% overlap off-set collision with a deformable barrier. In order to improve the predictive accuracy, a support vector regression model, based upon particle-swarm optimization, was used to simulate the relationship between output response and design variables. The overall result was that, following optimization, the weight of the structure was reduced and the crashworthiness was clearly improved. Lightweighting and crashworthiness aims can nevertheless sometimes conflict with one another. Tailor rolled blank structures, as noted elsewhere, are of potential value in reducing weight while improving crashworthiness. A study was made[53] of the crashworthiness of so-called top-hat structures, which were subjected to axial quasi-static/dynamic collision loadings, by experiment and numerical analysis. Three tailor rolled blank top-hat configurations, and corresponding conventional top-hat structures of uniform thickness were subjected to quasi-static/dynamic axial collision experiments. The tailor rolled blank structures exhibited the better crashworthiness; with more stable deformation occurring during the crushing process. Finite-element simulation results agreed well with the test data. The effects of the thickness distribution and the position of the transition zone upon the crashworthiness of the tailor rolled blank

Metallurgy of Lightweighting   Materials Research Forum LLC
Materials Research Foundations **133** (2022)   https://doi.org/10.21741/9781644902134

structures were such that they greatly influenced the crashworthiness of the latter. The results indicated that, for a given weight, the tailor rolled blank structures were better than the conventional structures with regard to the overall crashworthiness.

A space-constraint crash-box structure, when installed behind the fender, plays an important role in absorbing energy before transmitting it to the longitudinal rails. The crashworthiness of a multi-purpose crash box can be combined with lightweighting and ease of manufacture by choosing suitable cross-sectional shapes that connect to the usual crash rails. Numerical simulations can predict the energy-absorption ability and the mean load. In the case of a simple axial crush, a section of maximum effective width leads to a higher mean load and energy-absorption without damaging the longitudinal members. An important consideration is to restrain the crash-box tower from buckling and collapsing while allowing some intrusion. A reliability-based optimization model for frontal collisions was constructed[54] and an optimum polynomial model which was based upon error-ratio selection was applied to vehicle collision analysis. Probabilistic constraints were established by limiting the lower bound of the reliability interval in order to ensure the safety of the body structure. The results indicated that, following optimization, the total mass of the fender-beam, energy-absorbing boxes and front longitudinal beams was reduced by 2.4% while meeting all of the reliability constraints, thus ensuring safety while also achieving lightweighting. Because of the large deformations of a steel fender during low-speed collision, and the drive for vehicle-lightweighting, a new aluminium alloy fender was developed[55]. Finite-element models for the original steel fender and a newly-developed aluminium-alloy fender were established and three-point static loading simulations of the fenders were performed before testing. The results showed that the strength performance of the aluminium-alloy fender was better than that of the steel fender. Simulations and real sled-tests indicated that, under the same conditions, the deformation of the aluminium anti-collision beam and the deceleration of the sled were smaller than in the case of the steel fender. The crashworthiness of the aluminium-alloy fender was thus better than that of the original steel version.

Table 10. Properties of common engineering materials

Material	ρ (g/cm³)	UTS (MPa)	e (%)	E (GPa)	Specific Strength (MPacm³/g)
AZ91	1.82	280	8	45	154
AA6005	2.78	285	12	68	102
carbon-steel	7.86	517	22	200	66
Ti-6Al-4V	4.4	1100	20	110	250
ABS	1.03	35	40	2.1	34
PC	1.23	104	3	6.7	85
CFRP	1.5-1.7	-	-	131-200	~1000
GFRP	1.2-1.7	-	9.2-25	~166	

ABS: acrylonitrile butadiene styrene, PC: polycarbonate, CFRP: carbon-fiber reinforced plastic, GFRP: glass-fiber reinforced plastic

Another mandated safety feature is the seat-belt and its necessary anchorage. This leads to increases in weight and in the loads imposed on the seat. Simulations have been made[56] of safety tests of belt-anchorage points, showing that the maximum forward displacement at the upper anchorage point of a middle-seat safety-belt attained 433mm; thus failing to satisfy regulation requirements. A structural modification that was based upon force transmission was carried out which increased the seat mass by only 0.18kg, while the maximum forward displacement at the anchorage point was reduced to 104.3mm and the maximum strain was decreased from 0.56 to 0.163. When this single-seat modification was incorporated into a body-in-white, the simulation results indicated a maximum forward displacement and strain at the anchorage point of 220.9mm and 0.198, respectively. A proposed lightweighting scheme predicted a seat-weight reduction of 1.95kg, with the maximum forward displacement at the anchorage point increasing by only 10.4mm. Validation tests indicated that the simulation results agreed well with the test data, with the difference in the maximum forward displacement at the anchorage point being only 8.7mm.

Table 11. Relative cutting resistances and heat-of-fusion

Material	Cutting Resistance	ΔH_f (kJ/mol)
magnesium alloys	1.0	8.7
aluminium alloys	1.8	10.7
brass	2.3	13.1
cast iron	3.5	13.8
steel	>5	13.8

A computer study[57] of the lightweighting of a car yielded suitable thickness and property parameters for vehicle panels. Lightweighting of the car body yielded a 9.1kg reduction in body-weight; corresponding to a 5.44% lightweighting effect. The chassis is an ideal target for weight reduction and, when built using aluminium and magnesium alloys, this can yield weight reductions of 60% or 70%, respectively. This also reduces the overall stress and offers a higher factor of safety, but a drawback[58] is that lightening a vehicle body tends to increase noise and vibration.

Lighter-gauge panels tend to generate greater vibration and interior noise. This can be minimized by applying panel-damping treatments. A typical result is a 5.42dB reduction at the expense of a 1.37% increase in weight. Lighter-gauge panels are used to construct the monocoque structures which are the basic component of a vehicle. Because lighter-gauge panels tend to generate more vibration and interior noise, it is necessary to optimize the dynamic performance of lightweight vehicle structures so as to obtain acceptable levels of vibro-acoustic performance. The design of a light commercial van evolved over time and, due to lightweighting, could eventually be made some 10% lighter while increasing the volume by 15% and the load-carrying capacity by 18%. The overall performance was unsatisfactory however due to the occurrence of local resonant modes in the two side-panels. A finite-element model was developed[59] in order to determine the effect of adding stringers to the roof and side-panels so as to eliminate some of the local panel modes.

Conventional acoustic treatments tend to be very heavy, thus negating some of the weight-savings. It has been found that a lightweight sound treatment could be very effective against road noise, but less effective against engine noise. The addition of absorption treatments to the underbody of a vehicle helped to reduce road-noise, but was less effective with regard to engine noise. Lightweight parts, when working together, provided sound-absorption within the frequency range of interest. In the case of a heavy-duty truck tractor, the cast iron of the transmission shell was replaced[60] by aluminium. This weight-reduction led however to irrationalities in frequency-matching, and the vehicle-noise worsened. The genetic-algorithm method was used to optimize the power-train mounting system, and the mounting stiffness was chosen to be the design variable. The vibration energy distribution of the power-train mounting system was selected to be the objective of optimization. It was found that optimization of the power-train mounting system led to a marked improvement in vehicle vibration and noise.

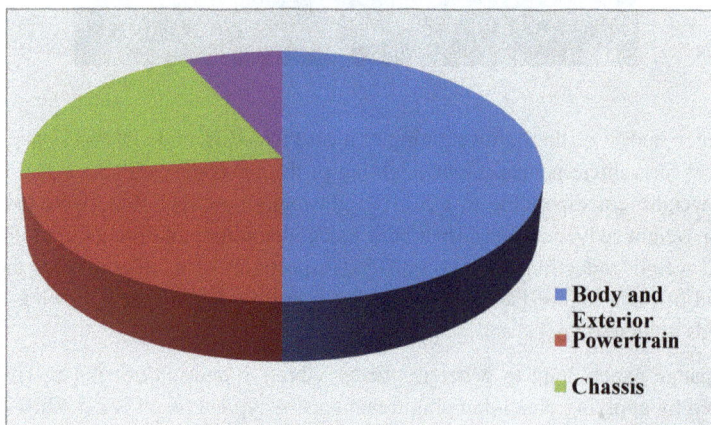

Figure 2. Sub-system weight as a fraction of curb weight for an average family car

A 19.0% decrease in the weight of a conventional vehicle can be achieved by using plastics and composites to replace both non-structural and structural components, with the seating offering the best target for weight reduction. The seating has to provide comfort as well as safety. The structural composite components are able to maintain the crashworthiness of the original vehicle in head-on and side-on collisions. An extended-range electric vehicle has been studied[61] with regard to the effect of key panels upon head-on crash safety. Such a mixed-materials approach requires a good understanding of the limitations and advantages of the available materials and of the possible joining methods[62]. For example, the use of aluminium and its alloys where the strength of complex shapes is required can meet weight-reduction targets without any loss in performance. Wrought aluminium alloys in particular offer marked weight-reductions of structural parts while preserving strength and safety[63]. Lightweight suspension components improve ride-quality, fuel consumption and emission reduction. The production of advanced high-strength steel and wrought aluminium produces average lifetime greenhouse-gas emissions of 3.9 and 17.5kg_{CO2}[eq]/kg, respectively[64]. Lightweighting by using the advanced steel to replace conventional steel resulted in lifetime savings, due mainly to savings in material use. Wrought aluminium had a much higher amount of associated greenhouse-gas production than did the advanced steel, and required greater fuel-economy to be exercised in the use-phase in order to achieve net savings. The maximum greenhouse-gas reduction occurred when the power-train was re-sized, when travel was congested or when the lifetime travel-distances were long.

An analytical solution to finding the optimum degree of lightweighting for a battery-driven electric vehicle has shown[65] that additional costs due to lightweighting can be off-set by the cost-savings arising from the smaller battery and motor which are required for a given performance and range. For a medium-sized vehicle, the optimum weight reduction[66] was of the order of 450kg, leading to an estimated manufacturing cost reduction of 4.9%. Smaller power-train costs were expected to decrease the importance of lightweighting in minimizing vehicle cost. So-called mass-decompounding is the identification of advantageous further secondary reductions. The most complex parts, which consequently require significant effort, are here designed first and are fixed while the rest of the development progresses. This maximizes the weight savings. A Monte Carlo simulation predicted[67] that the mean theoretical secondary weight-saving potential was 0.95kg per 1kg of primary weight reduction, with the possibility of 0.77kg per 1kg when all of the components were re-designed. When 4 key sub-systems among 13 were locked-in, the potential savings were 0.12kg/kg; with the possibility of attaining 0.1kg/kg.

Table 12. Mass influence coefficients for various sub-systems

Sub-System	Coefficient
suspension	0.14
body structure	0.12
motor	0.10
transmission	0.05
steering and braking systems	0.03
electrical system	0.03
tires and wheels	0.02
exterior	0.01

Vehicle lightweighting permits a decrease to be made in power-train size, and an appreciable reduction to be made in power-train cost. A method was proposed[68], for calculating the maximum net benefits, which considered both efficiency and powertrain-downsizing for a conventional internal combustion engine vehicle, for a battery electric vehicle with a range of 300 miles and for a fuel-cell electric vehicle. It was found that the neglect of powertrain-downsizing cost-savings undervalued the potential benefits of vehicle lightweighting; especially in the case of electric vehicles. As mentioned elsewhere, it is also important to limit the noise and vibration arising from the power-train. A lightweighting potential of 42kg was predicted[69] to reside in the power-train and chassis of a passenger car. The use of stronger transmission steels was expected to lead to a potential lightweighting of 2.45kg[70].

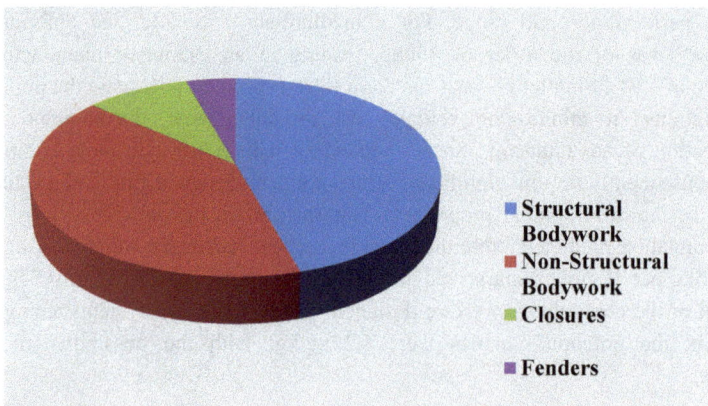

Figure 3. Sub-system weight as a fraction of body and exterior weight

The chassis, body paneling and power-train account for nearly all of the weight of the average family car. When extra weight is added to a vehicle, it has a series of knock-on effects in addition to the increased structural weight, such as an increased braking distance, raised fuel costs and a larger engine; if the previous acceleration is to be maintained. This is the mass-compounding effect which has to be counteracted. If the major components can be lightweighted, then other vehicle systems can also be lightweighted. If such changes cannot be made early in vehicle development, no appreciable weight decompounding is possible. Its effect can be characterized by using linear regression to define so-called mass-influence coefficients (table 12) which indicate to what extent a particular vehicle sub-system is dependent upon the vehicle weight. The

Materials Research Forum LLC
https://doi.org/10.21741/9781644902134

gross vehicle mass is the sum of those of the various sub-systems, the passengers and any cargo. The next step is to calculate what percentage of the components is mass-dependent, and some 66% of the total vehicle curb-mass (figure 2) is said to be dependent upon the gross vehicle mass. The remainder of the dependent mass is made up of the interior decor, dashboard, heating, air-conditioning, closures, fuel and exhaust system.

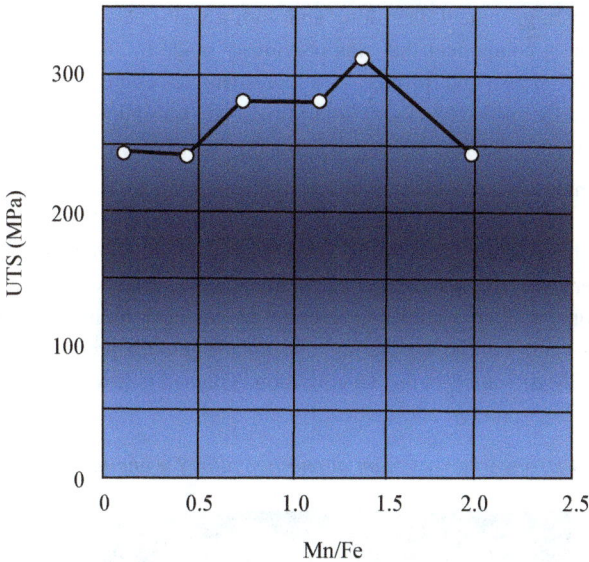

Figure 4. Ultimate tensile strength of aluminium-based engine-block liner

The mass-influence coefficient indicates that a 1kg reduction in vehicle weight results in a 0.14kg reduction in the suspension weight and a 0.51kg reduction in the overall weight. Mass-decompounding results, avalanche-like, in further weight-savings and to the definition of mass-decompounding coefficients. The mass-influence coefficients reveal that reducing the weight of non load-bearing sub-systems (figure 3) is important because secondary weight-savings in the body-paneling and power-train can be made. In the case of an average family car, a typical value for the sum of the mass-decompounding coefficients is 1.04; thus indicating that, if 1kg of weight can be avoided then about 1kg of additional weight-saving can be achieved. The use of lightweight sheet-metal can

Materials Research Forum LLC
https://doi.org/10.21741/9781644902134

reduce weight by 35 to 50% in the case of aluminium and by up to 60% in the case of magnesium. The use of these metals in other forms can unfortunately create formability and cost problems, although the formability problems can possibly be avoided by annealing during forming or between forming operations. The high-temperature ductility of magnesium in particular makes hot-forming feasible. On the other hand, the possibility of aluminium castings being made in the form of large hollow shapes but with a wall thickness of only about 4mm, makes aluminium suitable for manufacturing large chassis components. Smaller interior components can exploit the extreme fluidity of magnesium. A one-piece die-casting, some 150cm-long and with a wall-thickness of between 2 and 2.5mm, can replace a component that was previously made by the welding together of over 2 dozen steel parts, and results in a considerable weight-saving. The use of aluminium castings for engine blocks, cylinder heads and transmission housings has long provided a 50% weight-reduction in comparison with cast-iron use.

There is considerable unused opportunity for lightweighting by making careful casting choices. Casting is versatile, and some 22 of 24 cars were observed[71] to have aluminium wheels. Among all vehicles, 9 out of 10 wheels were aluminium and, the lighter the vehicle the larger was the advantage in weight-saving. The semi-solid casting of aluminium components is an established route to producing vehicle parts, but is under-used. A study was made[72] of the application of rheocasting to the production of an engine bracket from AlSi7Mg0.3 alloy in the T6 condition. This led to a weight-reduction when compared with the high-pressure die-cast equivalent.

Table 13. A strong wear-resistant aluminium alloy for engine-block lining

Element	Minimum (wt%)	Maximum (wt%)
silicon	9.5	12.5
iron	0.1	1.5
copper	1.5	4.5
manganese	0.2	3.0
magnesium	0.1	0.6
zinc	0	2.0
nickel	0	1.5
titanium	0	0.25
strontium	0	0.05

A traditional trick which is used in engineering in order to stiffen an angle-bracket and thus reduce the overall bracket weight is to use a dart. This has a nose-like appearance, is created in the corner of the bracket and is located at its middle. It also tends to reduce spring-back. It is interesting to note that, although darts have been a staple feature of metal-working for a very long time, the new and intense study of lightweighting has led to the realization that dart-design had been largely a matter of intuition and that careful theoretical and experimental studies had now to be performed in order to put dart-design on a more solid foundation[73]. Finite-element methods and simulations were compared with forming experiments, which largely validated the accuracy of the predictions.

It may also be possible to envisage making all, or most, of the remainder of the engine block out of magnesium alloy. The problem with aluminium engine blocks is that an iron liner is required, but new aluminium alloys may even be able to replace these. Such an alloy, which offers a suitable strength and wear-resistance, has been developed[74] (table 13). It is further specified that, when the iron content is greater than 0.4wt%, the manganese to iron weight ratio should be 1.2 to 1.75 or higher, and that, when the iron content is less than 0.4wt%, the manganese to iron weight ratio should be at least 0.6 to 1.2. The objective of this is to optimize the ultimate tensile strength (figure 4).

The reasoning employed here is that iron is habitually present in aluminium as an impurity, and material which contains less than 0.4wt% is particularly expensive. When the iron content is equal to, or greater than, 0.4wt% the weight ratio of manganese to iron should be between 1.2 and 1.75. When iron is present at a level of less than 0.4wt%, the weight ratio of manganese to iron should be between 0.6 and 1.2 if the manganese content is at least 0.2wt%. For the purpose of casting in general, the iron content should not be greater than 0.8wt% but, in the case of die-casting, the iron content can be as high as 1.5wt% so as to prevent the cast metal from sticking to the die. Copper and nickel affect the manganese content of the alloy. Nickel is not an intended constituent of the alloy but is acceptable in amounts of up to about 2wt%. Copper is also not an intended constituent, but does serve to impart strength. On the other hand, it is easier to avoid porosity (table 14) when the copper content is low. When the copper content is greater than 1.5wt%, or the nickel content is greater than 0.75wt%, it is preferable that the manganese content should be 1.2 to 1.5 times the iron content. Zinc is another common impurity but can be tolerated within limits. Titanium is introduced by scrap-recycling and has the effect of reducing the grain size when present to the extent of 0.04 to 0.25wt%. Strontium is added deliberately with the aim of modifying the eutectic aluminium-silicon phase in order to dendritic primary silicon-phase formation. Fluidity is the key parameter for evaluating the filling ability of casting alloys, and many factors affect the fluidity of aluminium-silicon casting alloys. Current understanding of the factors which influence the fluidity these casting alloys was systematically considered and the solidification mechanism of hypoeutectic aluminium-silicon alloys during high-pressure die-casting examined. It was concluded that this process is not inversely proportional to the solidification interval, but the fluidity-length increases with decreasing solidus temperature of the hypoeutectic alloys[75].

Table 14. Effect of various die-casting processes on aluminium porosity

Process	Die-Cavity Pressure (mbar)	Pore Volume (cm^3/100g)	Extra Cost
HPDC	1600	10-50	none
VADC	100-200	3-10	low/medium
SVDC	<50	<3	medium/high

HPDC: normal high-pressure die-casting, VADC: vacuum-assisted die-casting,
SVDC: super-vacuum die-casting

There has been a considerable Chinese interest in developing and patenting alloys for the specific purpose of automobile lightweighting. During the development and use of new advanced high-strength steels a key feature has been the use of a low-carbon material, micro-alloyed with niobium; examples being dual-phase, press-hardened complex-phase steels. One invention[76] was related to the smelting of niobium-containing phosphorus steel, from niobium-containing molten iron or niobium-containing pig iron, for the purpose of making high-strength steel plate. The smelting method involved high-temperature decarburization, vacuum deep decarburization and secondary refining. Another invention[77] was a novel copper alloy for automobile radiators. It was based upon oxygen-free copper with additions of phosphorus, tin and iron. The addition of tin greatly affected the softening temperature and thermal conductivity, while the addition of phosphorus had an additional de-oxidation effect and increased the softening temperature. The alloy overall was to contain 0.01 to 0.04%P, 0.05 to 0.3%Sn, 0.001 to 0.02%Fe and 78 to 90% of copper. It exhibited a high thermal conductivity, good corrosion-resistance and easy weldability. In the same vein, a new alloy was proposed[78] for automobile connecting-rods. This was a titanium alloy having niobium and silicon as additions. The effect of these additions was to improve the resistance to oxidation and creep. The novel material promised excellent mechanical properties and excellent corrosion resistance and was capable of reducing torque as well as aiding lightweighting. Another new alloy[79], for automobile wheel hubs, was magnesium-based, with silicon, manganese, molybdenum and iron additions. The material was described as being $MgMo_5Si_2Mn_2Fe$, and was to be die-cast. It had a high ultimate and yield tensile strength, good corrosion resistance, good thermal conductivity and good shock-absorbtion. It could efficiently radiate the heat arising from brake friction, and was easily recyclable as well as aiding lightweighting. A rare-earth containing aluminium alloy material promised[80] a high ductility and high strength, and had the composition: 2.0 to 3.5%Si, 3.5 to 5.0%Cu, 0.5 to 0.9%Mn, 0.1 to 0.5%RE, 0.1 to 0.2%Ti, 0.005 to 0.05%Sr and 80 to 92%Al. The RE component comprised cerium and lanthanum. The rare-earth elements were added in order to improve fluidity, reduce shrinkage and contraction-cavities and avoid the appearance of cracks. It was to be produced by indirect extrusion casting and was expected to be useful for automobile lightweighting.

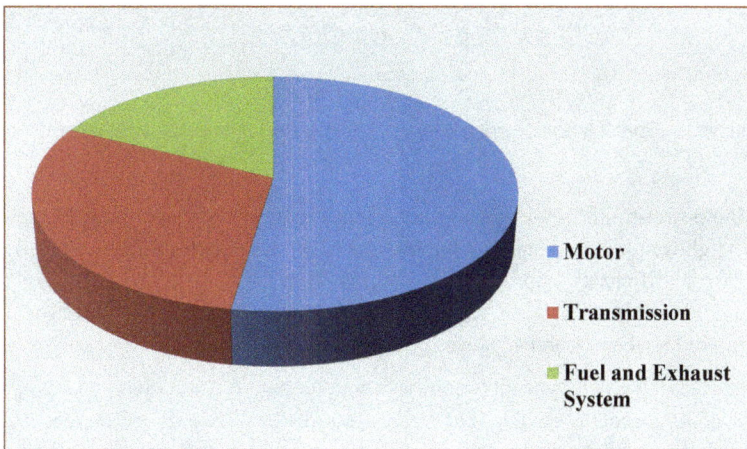

Figure 5. Sub-system weight as a fraction of power-train weight

The term, 'lightweighting', was first introduced[81] into the field of patenting in the context of reducing the weight of mirrors, although that concept already dated back at least to the construction of the Mount Palomar telescope.

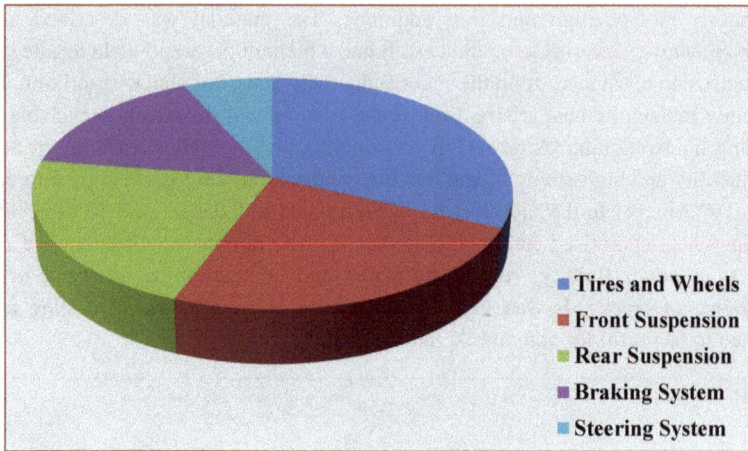

Figure 6. Sub-system weight as a fraction of chassis weight

Improving power-train (figures 5 and 6) efficiency reduces the scope for reducing energy use by lightweighting. That is, lighter vehicles benefit less from an efficiency improvement. Efficiency improvements ranging from 20 to 30% and lightweighting of between 200 and 600kg were optimum[82]. A reduction in the total cost of vehicle ownership of between 18 and 42% was possible.

A lighter electric vehicle typically requires a smaller battery for a given operating range, in turn leading to energy-saving. The vehicle and battery weights are pivotal factors in energy-efficiency. In the case of internal combustion propelled vehicles, any designed increase in the weight of a component has a 'knock-on' effect upon the rest of the vehicle in that other components have to be re-sized; thus further increasing the vehicle's weight and encouraging the aforementioned mass-decompounding (table 15). The same principle of course applies to battery electric vehicles: an initial weight-change results in secondary weight-changes and the reduced overall weight then reduces the required energy-capacity of the battery for a given operating range … and a yet-further reduction in the total vehicle weight.

A comparison[83] of a range of lightweighting options, which included all life-cycle stages, emphasized the importance of expanding the analysis beyond the use-phase, and predicted that the maximum achievable reductions in environmental impact were about 7%. Lightweighting strategies which were based upon the use of aluminium were found to be the most robust and consistent in reducing environmental impact. The benefits of using magnesium were less clear, and depended upon potential recycling. Carbon-fiber composites brought similar environmental benefits, to those of aluminium, but generally at a higher economic cost.

Table 15. De-compounding coefficients for various sub-systems

Sub-System	Coefficient
suspension	0.29
body structure	0.25
motor	0.21
transmission	0.10
steering and braking systems	0.07
electrical system	0.06
tires and wheels	0.04
exterior	0.04

Additive manufacturing and fused deposition modeling of effectively printed parts offer the possibility of producing parts which are hollow, sparsely-filled or have a lattice structure. Optimum configurations which both minimize weight and minimize printing-time can be identified.

All that the user needs to do is to provide is the current shape of a part, or the maximum space that the part can occupy, plus the maximum loads that it will experience. Advances in this relatively old technology now permit the rapid and inexpensive creation of previously 'impossible' parts. A disadvantage is that the magnitudes of the loads may be unknown before the design stage, and become known when it is too late for optimization. A modelling and simulation approach to the optimization of structures in the absence of any knowledge of service loads has been proposed[84]. This determines the best weight-saving while maintaining the same operational performance. The basis of the method is that optimization can be carried out by using relative loads, so that knowledge of the actual values is not necessary. This hypothesis has been successful, and has achieved weight-savings of up to 20%.

Studies of three-dimensional printed automotive fixtures exploit computational methods, topology-optimization and build constraints in order to manufacture a part. The general aim is to optimize the design, increase the stiffness of the original part and reduce material costs. One study[85] focussed on the topologically optimized design approach, using case studies of the lightweighting of aviation safety-critical parts. A particular area of interest for aviation light-weighting is the design of solar-powered aircraft wings[86]. Their high-aspect ratios tend to be associated with insufficient stiffness and this leads to non-linear deformation and other problems. The topology-optimization was carried out by using a so-called solid isotropic material with penalization algorithm in which discrete optimization problems are converted into continuous ones. The primary objective of the optimization studies is always to maximize the stiffness of the part while minimizing its mass. An investigation was also made of the effect of design constraints in order to account for feasible manufacture of the part while maintaining its structural response to loads. The optimized parts were then analyzed by using a lumped layer approach to simulate powder-bed fusion. The effect that the orientation of the build-direction has upon the result was subjected to parametric study.

The material extrusion family of additive manufacturing can however be very expensive for component fabrication due to the long production times for large complex components, and the associated cost of materials. The parts are typically built either as a solid in order to obtain the required strength or are built using a so-called shell technique in order to minimize the production time and the amount of material. The long-term aim is to introduce mechanical-strength requirements into an optimization model in order to determine the best fabrication route. Various internal structures such as orthogonal, hexagonal or interspersed pores have been considered in order to understand the mechanical-performance characteristics of various light-weighting configurations. It was found[87] that the bead deposition-path has an effect upon the mechanical characteristics, and that this factor should be taken into account.

An historical study showed[88] that, although the amount of low-density material in a vehicle has increased, the overall weight itself has also increased. This was attributed to the effect of consumer demand. When envisaging lightweighting, the substitution of conventional steel for high-strength steel is the most promising way of reducing greenhouse-gas emissions. The historical data showed that the number of greenhouse-gas payback-miles has increased, and will continue to increase, unless the reduction in greenhouse-gas emissions occurs at a higher rate than that of the fuel-efficiency improvement of steel-based vehicles.

Even the composite leaf-spring has become a key element in vehicle lightweighting[89] because of its better performance, when compared with a steel equivalent, even though its manufacture remains problematic because of cost and reliability issues. In order to design the connecting structures of the composite leaf-spring of a light bus, the ultimate loads suffered by the connection structures along various directions were deduced[90] by using multi-body dynamics simulations. A parameter-matching design methodology was proposed for choosing the connecting structure of a composite leaf-spring. Fatigue bench-tests and real-vehicle reliability tests of optimized connecting structures were performed, indicating that the chosen connecting structures for the composite leaf-spring could satisfy the requirements for its installation on vehicles. The development of a composite leaf-spring for a lightweight tractor demonstrated[91] the process of design and verification from the point-of-view of structural design, performance prediction and computer-aided engineering. Test results showed that, under conditions of equal stiffness, the weight and fatigue-life of the composite spring is more than 60% less, and over 3 times longer, as those of conventional steel-leaf springs, respectively.

A reduction in the weight of air-springs is another target of lightweighting. The problem of high local stresses in the air-spring bracket of a bus was tackled[92] by structural modification, guided by finite-element analysis and topological optimization. The least-squares method was applied to the curve-fitting of zig-zag boundaries in a topologically optimized model so as to make them smoother and more amenable to manufacture. Plug welding holes were also modified on the basis of the topological optimization, and the modified bracket was incorporated into a computer-aided engineering analysis. This showed that the maximum and average stress-levels in the modified structure were clearly lowered when the mass of the bracket was reduced by 18.6%.

In similar bracket-related work, topological optimization was carried out[93] by using a variable-density method. The geometrical model of the bracket was rebuilt and a finite-element analysis was carried out before comparing the stresses and deformations before and after optimization. This showed that the topology optimization method is effective in improving the mechanical loading response and in aiding lightweighting. Topological optimization has also been applied[94] to a vehicle rear sub-frame[95] by constructing an implicit parametric model of the frame and applying a cross-sectional shape-control which took account of 36 parameters, including part-shape, position, thickness and material. Multi-objective optimization was then carried out on the topologically optimized model, with mass minimization and first-order modal-frequency maximization as objectives and hard-point stiffness plus the first three-order modal frequencies as constraints. The results showed that the rear sub-frame weight could be reduced by 2.41kg; a lightweighting of 14.5%. Multi-material topology optimization has proved to be a valuable approach to making the optimum choice of material type and distribution within a complex vehicle structure. One can envisage 1-material, 2-material and 3-

material optimization, and determine the effect of the number of materials upon the design and its predicted performance. Individual material combinations and their effect upon the final design performance can be explored. In effect, the method is a direct extension of the classic single-material problem. In some versions, the topological optimization problem is expanded so as to include material selection variables in the usual density method while using a solid isotropic material with penalization interpolation. The key features include interpolation schemes, sensitivity analysis and filtering methods.

Aluminium

The replacement of denser metals by aluminium is an attractive option for weight reduction. The use of aluminium (tables 16 and 17) in vehicles has increased significantly for decades as a result of its utility in power-trains and suspensions as well as in the lightweighting of body structures. The aluminium also becomes available for recycling at the end of vehicle life, and some 90% is recovered[96]. Since the early days of modern metallurgy, commercial aluminium alloys have been developed for use within a particular range of applications for which the properties do not necessarily coincide with those required for lightweighting. This has led to new alloy-design philosophies which yield an expanded profile of properties that offer a superior compromise between formability and strength. One additional target is to maximise the possibility of recycling any resultant articles. These new so-called cross-over alloys[97] combine the various desirable properties which are normally limited to specific classes of commercial aluminium alloy, and include those of the types, AlMg (5xxx), AlCuMg (2xxx) and AlZnMgCu (7xxx), which can offer good formability combined with a high age-hardening potential.

Aluminium/nylon laminates have been designed[98] for lightweighting in stiffness-limited vehicle body-panel applications. They were stiff and damage-resistant and appeared to be an adequate cost-effective replacement for 0.81mm steel sheet, while being two-thirds lighter in weight per square foot.

The possibility of using an entirely aluminium body to reduce vehicle weight has been considered[99], although there were concerns that were related to occupant safety. Attention was therefore paid to the energy-absorbing properties of aluminium components. It was concluded that correct alloy selection and vehicle design could result in a more mass-efficient energy-absorbing structure that would provide at least the same level of occupant safety as that previously offered by heavier vehicles.

Table 16. Predicted effect of various weight-reduction strategies

Materials	Status quo	Mainly steel reduction	Mainly aluminium reduction
high-strength steel (kg)	291	486	280 (97*)
cast iron (kg)	117	116	109 (50*)
other steels (kg)	589	341	327 (255*)
wrought aluminium (kg)	42	42	136 (310*)
cast aluminium (kg)	117	117	153 (181*)

* with maximum aluminium replacement

A major application of aluminium in vehicles involves cylinder heads, where it is valuable because of its low density and high thermal conductivity. Because of the ongoing increases in combustion pressures and temperatures, material properties are a key factor in the performance of aluminium cylinder heads.

Table 17. Properties of common aluminium alloys

Alloy	ρ (g/cm^3)	UTS (MPa)	YS (MPa)	e (%)	E (GPa)	Specific Strength (MPacm^3/g)
A380	2.7	315	160	3	71	117
AA6005	2.78	285	250	12	68	102
AA6082	2.69	320	260	12	68	119
AA5083	2.68	270	110	12	69	101
AA7020	2.82	350	280	10	70	124

Interactions were considered[100], between life-cycle emissions and material usage, which were associated with the lightweighting of vehicles. Both aluminium and high-strength steels were involved, with the degree of lightweighting ranging from 6% to 23%. The increase in greenhouse-gas emissions which was associated with the production of lightweight vehicles was compared with the degree of reduction in emissions which occurred during vehicle usage. This introduced the concept of calculating just how many years of vehicle use would be required in order to off-set the additional greenhouse-gas emissions which were generated during production. The number of years which was required in the case of high-strength steels was less than that for aluminium. The achievement of appreciable lightweighting, by using high-strength steel rather than aluminium, required not only material substitution but also the achievement of a secondary lightweighting effect by downsizing other vehicle components in addition to the vehicle structure itself. The amount of greenhouse-gas reduction in the case of aluminium-lightweighting depended markedly upon where the aluminium was produced and whether recycled aluminium was used. High-strength steels were less sensitive to such factors. The greenhouse-gas offset time which was required for vehicles that were lightweighted by using aluminium could be shortened by means of the closed-loop recycling of wrought aluminium. It was considered to be unlikely however that, on a 15-year time-scale, this could significantly reduce the emissions which arise from the automobile industry. Associated estimates were made[101] of the cost per metric ton which was involved in removing greenhouse-gas emissions by the use of lightweighting, and a distinction was drawn between additional manufacturing costs and fuel-cost savings which resulted from lightweighting. It was shown that greater greenhouse-gas reductions arose from greater lightweighting and added to manufacturing costs. The associated production costs were disproportionately higher than the fuel-cost savings that were associated with greater lightweighting. Vehicle-lightweighting was more cost-effective for larger vehicles.

Primary energy demand is reported as a measure of the energy efficiency of the whole system, from cradle to grave. The results generated for the full life-cycle of a B-pillar, indicated that a new design which involved DP800 and DP1000 steels had a lower impact on 4 standard impact categories assessed, as well as on primary energy demands. The effect of replacing one B-pillar was compared, for petroleum and diesel drive-trains, by separately considering manufacturing, use and end-of-life aspects. The results showed that the use-phase is the main contributor to the life-cycle impact of the B-pillar choice. Upon assuming a 200000km vehicle lifetime, and a petroleum drive-train, the use-phase accounted for 93% of the life-cycle global-warming potential of the DP800/DP1000 B-pillar option, and 94% of the global-warming potential of the boron-steel option. For each aspect considered, the use-phase accounted for more than 80% of the total impact. The reduction in fuel consumption which resulted from the 4kg weight-saving, and an associated power-train resizing, accounted for the most of the difference in impact of the alternative B-pillar designs. Analysis of the possibility of end-of-life steel-recycling revealed the positive effect of environmental savings during the next life-cycle due to the reduced need for steel extraction from ore. The comparison of two B-pillar designs showed that the new design also had a lower environmental impact. The global warming potential could be further divided into the effect of manufacturing alone, and the contribution made by molybdenum and other alloying elements. The small increase in environmental impact which was due to the greater use of elements such as molybdenum in the DP800 and DP1000 steels was more than compensated by the reduction in material required for the new B-pillar, and the associated reduction in the effect of iron-making and subsequent steel-making. Both types of B-pillar were assumed to be produced by using conventional forming processes. In order to improve the crashworthiness of a B-pillar for side-impact and simultaneously reduce its mass, the safety performance of a car was firstly analyzed[102] according to the design parameters of the B-pillar. It was found that the intrusion and intrusion velocities at the middle of the B-pillar were too high and that the mass of the B-pillar structure was also rather high due to extra reinforcements. An improved design of B-pillar was introduced by using the tailor-welded blank structure, and the location of weld and the thickness of the structure were optimized. The desired deformation mode of the B-pillar was finally deduced, giving a maximum intrusion reduction of 10%. The intrusion and intrusion velocities at the middle of the B-pillar were reduced by 18% and 12%, respectively. The mass was reduced by 18%. The optimum design of a B-pillar structure with tailor-welded blank could thus effectively balance the requirements of crashworthiness and light-weighting.

A third sensitivity analysis investigated the effect of using fuel reduction values without powertrain resizing. Reduction values without powertrain resizing are estimated to be $0.15l/(100\text{km}100\text{kg})$ for gasoline vehicles and $0.12l/(100\text{km}100\text{kg})$ for diesel vehicles. Reduction values with powertrain resizing were assumed to be $0.35l/(100\text{km}100\text{kg})$ for gasoline vehicles and $0.28l/(100\text{km}100\text{kg})$ for diesel vehicles. This sensitivity analysis showed that the reduction in impacts was 13–14% without powertrain resizing compared to 28–30 % with powertrain resizing. This illustrates that the DP800/DP1000 option is still likely to result in a lower environmental impact than the boron-steel option but that much greater benefits are likely to be possible with simultaneous powertrain-resizing. Other ways of affecting the use-phase include changing the total distance driven by the vehicle and modifying the driving cycle. Such innovations are likely to be the same for both types of vehicle and could thus potentially affect the proportion of total impact that could be attributed to the use-phase. On the other hand, they would be unlikely to produce large changes in the overall resultant trends. Results suggest that a 4kg weight-reduction leads to considerable use-phase savings and constitutes the overall difference in impact of the two components. The global-warming potential saving for both B-pillars in a popular family vehicle over 200000 km of total driving distance was estimated to be 165kg of CO_2[eq] for a petrol drive-train and 141kg CO_2[eq] for a diesel drive-train. This would be equivalent to driving the vehicle for 1000km assuming a standard driving cycle.

Cast Iron

Material replacement is complicated because not all materials can meet required specifications or are too environmentally and/or economically costly. Aluminium is an obvious and popular choice but the side-effects associated with its production are greater than those for denser metals such as iron and steel. When compared with iron, it has inferior mechanical properties with respect to strength, stiffness, high-temperature resistance and damping capacity. It is also more expensive. One way in which to make cast iron more competitive with cast aluminium is to reduce the thickness of cast iron products: popularly known as thin-walling. This can make cast iron products effectively low-weight while retaining its superior mechanical properties. It is now possible to produce thin-wall ductile cast iron items with cross-sections of less than 2mm. The traditional problems which were associated with castings, such as manufacturing difficulties and defects in thin walls, have now been largely overcome by altering the alloy composition, the processing temperature and using inoculation techniques.

Methods for predicting the defects which are introduced into castings during manufacture are needed. Cost and weight savings are already possible by using casting, but there is significant room for improvement. Advanced techniques have the potential to play an important part in the optimisation of castings but accurate material property values are essential. The latter must include local defects or variations in properties that are introduced during manufacture.

Ductile-iron castings offer considerable opportunities with regard to lightweighting. As noted, ductile iron castings which are free from massive carbides can currently be produced with dimensions of less than 2mm by making changes in their composition and processing. If an aluminium component is more than 4mm thick, its replacement by ductile iron can achieve both lightweighting and reduced cost[103]. One of the key differences between conventional and thin-wall ductile iron castings is the composition. Varying the amount of pig iron and alloying elements such as silicon helps to control the alloy properties (e.g., tensile strength, nodularity, hardness, castability, ductility). Variations in composition influence the production-impact of a given alloy because of the relatively higher material impacts of materials like pig iron and silicon as compared with that of scrap iron. It is therefore important to determine whether the reductions in fuel consumption due to lightweighting are sufficient to offset the possible increase in the production impacts of alternative alloys. Life-cycle assessment is the method of choice for evaluating the performance of products over their lifetime, and it has found extensive use in the automotive industry. There is much existing information on thin-wall ductile iron casting in general, but information on its life-cycle performance is somewhat lacking. A parametric model can assess the life-cycle performance of thin-wall ductile cast iron as compared with conventional cast iron and cast aluminium. Attention should be paid to the lightweighting of specific components and production, use and end-of-life handling of both conventional and thin-wall components should be calculated. Further analysis should then determine the break-even point between increased production-phase impact (including the use of pig iron) and decreased use-phase impacts by varying the assumed amount of lightweighting. Assessment of the life-cycle performance also involves considering all of the life-cycle phases, including material sourcing, manufacturing, use and end-of-life; always including estimates of energy expenditure and greenhouse-gas emissions. The casting method is assumed to be the same for both thin-wall and conventional cast iron and thus the main difference is that of alloy composition and the actual amount of material required.

The choice of light high-strength vehicle materials lies between advanced high-strength steel (figure 7) on the one hand, and aluminium metal-matrix composites, magnesium or polymers on the other[104]. Differing views of life-cycle assessment use for judging sustainability in the aluminium industry have led however to very different results[105]. For example, the greenhouse-gas emissions per kilogram of primary aluminium production range from 5.92 to $41.10 kg_{CO2}$[eq]. The point at which the fuel-economy benefits of lighter vehicle off-set the additional emissions resulting from production, have been estimated to range from 50000 to 250000km. A study was made[106] of the potential life-cycle impact of the alternative lightweighting material solution designs involving steel and aluminium. Life-cycle assessments were made of three lightweight-vehicle designs: current (baseline), advanced high-strength steel and aluminium, and aluminium-intensive. Secondary weight-savings which resulted from body-lightweighting were considered with regard to the engine, powertrain and suspension. Designing a lightweight powertrain mount within a given packaging space while satisfying performance requirements is quite a difficult task and requires the use of techniques which are beyond usual methods of design synthesis.

The life-cycle assessment included material production, semi-fabrication, use-phase operation and end-of-life recycling. The use-phase weight reduction was found to account for over 90% of the total vehicle life-cycle energy and CO_2[eq] emissions. The aluminium-intensive design yielded a weight reduction of 25% with respect to the baseline design, resulting in reductions in the total life-cycle primary energy consumption of 20% and a reduction in CO_2[eq] emissions of 17%. The aluminium-intensive design promised the best break-even vehicle mileage, from the point-of-view of both primary energy consumption and climate change benefits.

▪	HSLA450, BH340, 400 (32.7%)
▪	DP500,600 (11.8%)
▪	HF1500 (11.1%)
▪	DP1000 (10%)
▪	DP800 (9.5%)
▪	TRIP780 (9.5%)
▪	CP1000-1470 (9.3%)
▪	Mild Steel (2.6%)
▪	TWIP980 (2.3%)
▪	MS1200 (1.3%)

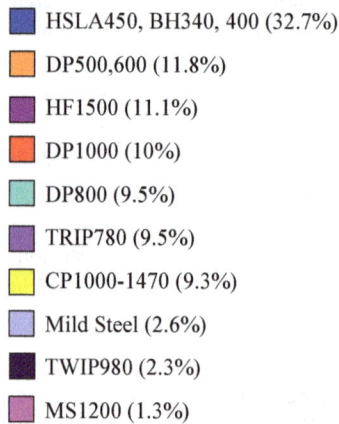

Figure 7. Advanced high-strength steels in electric vehicles: contribution of various steel grades to cars designed for the Future Steel Vehicle (FSV) program. Reproduced under Creative Commons Attribution Licence (with modified key)[107]

Aluminium-silicon alloys are staples of the aluminium-foundry industry, and are cast at rates in the millions of tonnes per year, but their coarse microstructures limit the mechanical properties and thus their utility in vehicle-lightweighting. A new family of cast aluminium-silicon alloys has been found which produce *in situ* nano-composites that contain up to 25vol% of ultrafine equiaxed silicon particles within an aluminium-alloy matrix. The alloys could be ductile, or reinforced by nano-scale spinodal constituents. Hypereutectic A390 alloy was solidified, solution-treated and artificially aged[108] in order to spheroidize the silicon and dissolve intermetallics so as to re-precipitate them in the solid state as nano-sized spinodal structures. The nano-scale structures[109] of the new materials imparted a markedly improved strength, hardness and wear resistance, together with adequate toughness and ductility. The new materials could be prepared, using continuous strip-casting or high-pressure die-casting, from conventional low-cost aluminium-silicon melts while offering a much higher volume fraction of ultrafine silicon dispersoids than was expected of *in situ* formed materials.

The development of high-performance aluminium-silicon cast alloys is important with regard to vehicle-lightweighting. The effects of silicon (5% to 7%), copper (1% to 3%) and magnesium (0.3% to 0.9%) upon the mechanical properties of Al-Si-Cu-Mg cast alloys (table 18) were studied[110], showing that an Al-6Si-3Cu-0.3%Mg alloy exhibited better overall mechanical properties after a T6 heat treatment (table 19).

Table 18. Properties of as-cast Al-Si-Cu-Mg alloy

Si (wt%)	Cu (wt%)	Mg (wt%)	UTS (MPa)	E (%)
5	1	0.3	192.12	5.29
5	2	0.6	202.26	2.92
5	3	0.9	219.22	2.33
6	1	0.6	216.08	4.68
6	2	0.9	232.5	3.30
6	3	0.3	188.76	4.31
7	1	0.9	232.93	4.60
7	2	0.3	239.19	5.52
7	3	0.6	229.53	3.58

An aluminium-alloy front fender beam for a vehicle was designed[111] which was 35% lighter in weight than the existing DP1200 ultra high-strength steel version. During a head-on collision at 25km/h, the first acceleration-peak for the aluminium version was 2.6G higher and occurred 5ms earlier, as compared with the equivalent data for the steel version. This implied that the aluminium-alloy fender could better protect the occupants. Simulations showed that, in a head-on collision at 10km/h, the maximum intrusion for the aluminium-alloy fender beam was 7mm less, and the energy-absorption was 70% more, as compared with the equivalent figures for the steel fender beam.

Table 19. Properties of Al-Si-Cu-Mg alloy in the T6 condition

Si (wt%)	Cu (wt%)	Mg (wt%)	UTS (MPa)	E (%)
5	1	0.3	276.08	4.16
5	2	0.6	301.61	2.65
5	3	0.9	282.43	2.06
6	1	0.6	320.33	4.55
6	2	0.9	347.6	3.04
6	3	0.3	382.01	3.82
7	1	0.9	343.48	3.45
7	2	0.3	325.34	3.63
7	3	0.6	369.93	2.45

In order to maximize the value of lightweighting, replacement with lighter materials has to extend down to every detail. This can involve, for example, replacing a gear-shift made from cast iron with one made from die-cast aluminium, without of course sacrificing strength or rigidity. Die-cast aluminium possesses good mechanical and thermal properties as compared with those of cast iron.

An important factor in weight-reduction is a concomitant decrease in the powertrain weight. A reduction from 340 to 267kg can permit the use of a 1-liter 3-cylinder engine as the main powerplant. By down-sizing the engine and changing the materials used in its construction, the weight of the engine might be lowered by 29kg. Use of an aluminium engine-block with steel bulkhead inserts, and a connecting-rod forged from high-strength aluminium alloy, can yield appreciable weight-reductions in comparison with the use of equivalent conventional iron and steel components. On the other hand, when an aluminium alloy is used it might itself not be able to support the loads associated with operation, and thus require reinforcement. A cast aluminium engine-block with reinforcement could still reduce the overall engine weight. High-temperature high strain-rate forging can improve the ductility and fatigue strength of 2618-T6 aluminium alloy to such an extent that it can be used in a connecting rod that satisfies fatigue safety factor requirements. Such an alloy might reduce the weight of a connecting rod by over 200g; a weight-saving of 40% when compared with a conventional steel alloy rod. Aluminium-alloy engine blocks certainly aid lightweighting, but create problems of thermo-mechanical mismatch between the aluminium and grey cast iron cylinder-liners, which lead to high tensile residual stresses appearing along the cylinder bores. The mechanical properties in that region then have to be improved in order to prevent premature engine-failure. Billet castings have been used[112] to simulate the engine-block solution heat-treatment and to determine the onset of incipient melting. The results indicated that solution heat-treatment at 500C dissolved secondary-phase particles, while solutionizing at 515 or 530C led to incipient melting of Al_2Cu and $Al_5Mg_8Cu_2Si_6$. Incipient melting in turn led to the formation of ultra-fine eutectic clusters which consisted of aluminium, Al_2Cu and $Al_5Mg_8Cu_2Si_6$ upon quenching. Incipient melting began at 507C in all of the billets, but the degree of local melting decreased with increasing microstructural refinement, as deduced from lower endothermic peaks and energy-absorption.

The use[113] of an aluminium cross-beam in an independent front-suspension led to a weight-saving of 40 to 50% when replacing welded-sheet structures. It also offered an improved corrosion resistance and the possibility of profitable recycling[114].

A novel design was developed[115], for a front rail, which maximized collision energy-absorption. Simulations of the extrusion crush behavior were performed using anisotropic yield functions, and the simulation results were compared with dynamic crush results for the extrusion. The size of the structure was optimized by using response-surface methodology, artificial neural-network metamodels and simulated annealing. The specific energy absorption was used as a target parameter when maximizing energy-absorption and minimizing weight. An analytical relationship was found which related the specific energy absorption to the crush efficiency, thus showing that a single optimization parameter could be sufficient for targeting weight-minimization. Further analysis identified key extrusion parameters, and the wall-thickness was found to be the most important parameter to be controlled during extrusion.

Attempting to combine lightweighting with a high specific energy absorption makes aluminium foam an ideal energy-absorbing material. The contribution made by aluminium foam composite structures to absorbing collision-energy has been explored[116]. A model aluminium-foam composite structure was first built, with test data providing the basic data for simulations. The energy-absorbing effects were then verified for the case of a vehicle sill-crossbeam. This showed that an optimum design could markedly lower acceleration and reduce intrusion.

High-strength 7075 aluminium alloy is an ideal replacement for steel when lightweighting vehicles, but it suffers from inhomogeneous structures and marked segregation when conventionally cast. A study has been made[117] of slurry preparation of the alloy by means of spiral electromagnetic stirring and rheo-diecasting. Using experimental and simulated data, the shape-filling and solidification of engine connecting-rods made of 7075 alloy were optimized. The computational model which was used described well the rheo-diecasting filling and solidification process, and an acceptable connecting-rod was obtained at an extrusion-rate of 3m/s and mould temperature of 150C. The microstructures and mechanical properties of rods which had been manufactured using rheo-diecasting and normal die-casting were compared, showing that the average grain-sizes of rheo-diecast and normal die-cast rod were 34 and 59μm, respectively. In the case of rheo-diecasting, the microstructure was finer and more uniform, and the segregation of zinc, magnesium and copper was markedly reduced. The hardness was also 19HB higher.

Superplastic forming of aluminium alloys has been used to create vehicle outer body panels and closures. Because such forming involves handling of aluminium sheet at higher temperatures, the final strength of the material is similar to that of annealed aluminium sheet. Current vehicle applications favour the use of AA5083 alloy which, given its relatively high magnesium content, has a post-formed strength of about 150MPa. The use of AA5000 series alloys having magnesium contents above 3% risks the occurrence of stress corrosion cracking during exposure to a high-temperature environment. The result is that superplastic AA5083 alloy has had restricted use in body-in-white and under-hood applications, where a higher strength and resistance to stress corrosion cracking are required. The AA5000, AA6000 and AA7000 series aluminium alloys have been examined in order to produce a high-strength superplastic-forming aluminium sheet having a post-forming strength of above 250MPa. Modifications to existing heat-treatable alloys have been combined with thermomechanical processing methods in order to produce a fine-grained superplastic sheet having a post-formation heat-treatment response. The development of a fine-grained microstructure that remains stable at superplastic forming temperatures usually requires a combination of composition adjustment combined with combinations of hot- and cold-rolling. In the aluminium system, known alloy specifications permit wide ranges of many of the main alloying elements. Alloy modification within these alloy ranges can optimize superplastic formation microstructures. The aim is to develop a heat-treatable alloy which has a post-forming age-strengthening response that is compatible with vehicle body-in-white thermal treatments. The intended result is a high-strength superplastic aluminium alloy sheet which can be used in higher-strength and high-temperature vehicular applications, thus extending the use of superplastic forming of aluminium to body-in-white and under-hood components. New melt additions for the control of residuals in secondary aluminium have been developed.

High-formability aluminium sheets having an equal formability have been compared to high-tensile steel sheets by considering the Lankford value. This is the ratio of the sheet-width strain to the sheet-thickness strain when a tensile stress is applied. The drawability is higher when the Lankford value is higher. The aim is to achieve this maintaining the current tensile strength and elongation. The Lankford value depends greatly upon the texture, and the latter needs to be controlled to a high degree in order to increase its value. The Lankford value is intended to be improved by thermally stabilizing the aluminium rolled texture to give a high Lankford value during rolling and by carefully controlling texture so as to form a recrystallization texture which is similar to that of steel sheets. The average Lankford value was 1.3 for constant-temperature rolling. Asymmetrical warm-rolling was selected as an alternative. Sheets which were rolled by asymmetrical warm-rolling exceeded had a Lankford value greater than 0.9. An intermediate target was a Lankford value of 1.1, with the final target being 1.2. The drawability was also improved as compared with that of conventional alloys. Further targets are a higher Lankford value and the establishment of stable rolling conditions. Another aim is to use aluminium to replace steel bodies, especially in B-pillars, side-sills and front-side members. Technology needs to be developed in order to join steel and aluminium as to build monolithic structures from aluminium and steel. Spot-welding and arc-welding can join almost all steel-steel structures. Hybrid structures can be manufactured without greatly modifying existing production lines, if those same joining methods can be used. Spot-welding and arc-brazing methods achieve equal joint strength to that of an aluminium alloy with regard to both shear strength and peel strength.

When compressed, porous aluminium exhibits a unique load-displacement curve that cannot be matched by ordinary materials of hollow form. The compressive deformation behavior of porous aluminium is generally low under an initial maximum momentary stress. The stress exhibits a plateau region which has an almost constant length; beginning in the elastic deformation region and leading to a densification region that is caused by a reduction in the number of pores. This characteristic deformation behavior explains the shock-protection ability of porous aluminium. It has been used as a shock-absorption material for automobile collision safety in the form of crash boxes and fenders, instead of hollow steel structures, hollow aluminium extrusions and the similar options chosen for conventional crash boxes. Displacement-load curves during crash tests show however that the initial loading experienced during deformation is large, and that large load-fluctuations occur during subsequent stages of the crash deformation process, such that the crash energy cannot be absorbed smoothly. Conventional crash boxes are therefore not ideally designed for passenger, pedestrians and car-body protection. The targets of an intermediate energy-absorption figure of 5kJ/kg and a final figure of 8kJ/kg have been achieved; thus exceeding the absorption energy of previous steel crash boxes and promising further weight reduction. Crash boxes and hollow frame members which are filled with porous aluminium exhibit especially interesting effects with regard to crash shock handling and bending stiffness. Exploitation of these features of porous aluminium promises simplification, small cross-sections, weight reduction, reduced cross-sections and the elimination of reinforcing members. Small cross-sections and miniaturization then free-up space for making other improvements in design. Further innovations in melt treatments and powder metallurgy can conceivably optimise bubble formation and continuous mould foaming technology.

Conventional porous aluminium production by melting is based upon pure aluminium and leads to a density of 200 to 300kg/m³ and an energy-absorption of 2 to 3kJ/kg. The density fluctuates widely but bubble-diameter control can improve the characteristics. As a result, the above intermediate target of 5kJ/kg could be achieved via density control ranging from low-density metal (low strength) to high-density metal (high strength). The main control parameter here was the bubble diameter. Parts which have a complicated shape, and those made from porous aluminium by using conventional melting methods, can be manufactured by first preparing large blocks by using batch production and then cutting and shaving them. Traditional production methods had encountered problems of low yield and low efficiency during machining.

When considering the use of powder metallurgy materials such as alloys of aluminium, titanium, and magnesium, each choice presents problems when manufacturing components in high volumes for cost-sensitive applications[118]. The Al-4%Cu alloy has been the most widely-used powder metallurgy material for over two decades in the form of automotive camshaft caps. This alloy offers yield strength and ultimate tensile strengths of close to 160 and 200MPa, respectively. Heat-treatment permits the attainment of higher strengths, but with the risk of embrittlement. The aim is to produce components having yield strengths of over 300MPa together with reasonable ductility. Such components can replace wrought, cast and powder-metallurgy steel components, giving a reduction in mass of at least 50%.

Materials Research Forum LLC
https://doi.org/10.21741/9781644902134

Mould-foaming technology permits the direct manufacture of complicated three-dimensional shapes. Continuous-foaming technology allows the production of members of simple shapes (sheets, rods) rather than having to cut them from blocks. Manufacture of foamed material of various shapes is possible by using a powder precursor method which involves heating the unfoamed precursor in a mould. By optimizing the heating process the porosity can be freely selected,, with a range of 80% or more in porosity and a range of 2 to 8mm in pore size.

Putting porous material into an aluminium mould increases energy absorption and a synergistic effect can be achieved by bonding together the mould and the porous material, imparting an energy absorption of 14kJ/kg. Improvements in bending stiffness and bending strength result from filling a hollow member with porous material. This made it possible to use smaller cross-sections and to simplify their shape and structure. Ideal structures for porous structures, with regard to pore shape and cell-wall properties, have been studied using three-dimensional computer tomography and optimised by means of finite-element analysis. Porous aluminium shapes can also be integrated into structures by means of welding and other joining technologies.

By using aluminium alloys instead of steel, a direct reduction in the weight of a vehicle of up to 47% can be achieved. Among the various aluminium alloys, A206 alloy has been widely utilized in the automotive industry due to its high tensile strength, yield strength and good fracture toughness. The evaluation of the high-temperature creep behaviour of the aluminium alloys which are used in the automotive industry has attracted much general attention, but the high-temperature tertiary creep stage of aluminium alloys has been relatively neglected. *In situ* neutron diffraction studies were made of A206 aluminium alloy under tensile loading at 225C such that the conditions simulated in-service conditions for automotive power-train components like the cylinder block and the engine head.

Neutron diffraction offers the advantage that neutrons are unaffected by the electron clouds of atoms, thus allowing the neutrons to penetrate deeper into a material. The specimens were machined from T-7 heat-treated cast bars of A206 and, before applying the load, the specimens were stabilized in air at 225C for some 200h in order to allow any non creep-related microstructural changes to occur before testing. Any overall strains which might be caused by thermal expansion and plastic-elastic deformation were investigated. The *in situ* neutron diffraction results revealed the tertiary creep behaviour of selected {111} planes within the A206 by monitoring the full-width at half-maximum of the diffraction peak and the creep strain. This innovative technique permitted the observation of damage-accumulation during tertiary creep up to the moment of actual fracture of the material at high temperatures.

An assessment of primary and secondary (steady-state) creep in Al–Si and A206 had already been made by means of *in situ* neutron diffraction analysis. As compared with secondary creep, the peaks during tertiary creep broadened considerably; indicating the presence of type-II and type-III stresses in the material. The type-II and type-III residual stresses resulted from the accumulation of severe intergranular plastic deformation. This created multiple zones of localized tensile and compressive forces which acting upon individual grains; that is, damage accumulation. It was in turn the direct result of severe distortion of the material occurring during tertiary creep. Profiles of the creep strain and full-width at half-maximum at the selected {111} planes were obtained during increasing cyclic loading. The full-width at half-maximum fluctuated slightly during steady-state creep under an applied load of 120MPa. The fluctuation was suggested to be a direct result of the dynamic balance between strain-hardening and dynamic recovery during deformation. After establishing steady-state creep, the load was lowered to 5MPa, and then increased again to 140MPa. The higher load then initiated tertiary creep. When compared with the steady-state creep secondary creep, the full-width at half-maximum increased markedly during tertiary creep. This increase in the full-width at half-maximum was a direct sign of distortion of the crystal lattice resulting from severe mechanical deformation of the material during tertiary creep stage and indicated damage accumulation before failure. Excessive deformation caused intergranular misalignment accumulation during the final stages of and resulted in excessive intergranular stress, micro-crack nucleation and growth and final failure. Subjecting materials to such *in situ* tests can permit direct assessment of the service-fitness of new materials for lightweighting critical components such as automotive power-trains. Assessment might also be performed *post facto* by assessing the full-width at half-maximum following severe deformation or impact even when strain data are not available. Assessment of the full-width at half-maximum via neutron diffraction extends the possibilities of such assessment.

A test-case study compared the environmental sustainability of a lightweight high-pressure die-cast primary aluminium suspension cross-beam, for light vehicles, with a conventional steel-sheet suspension cross-beam. The latter weighed 30kg while the former weighed 15kg, thanks to the introduction of a new design. A cradle-to-grave life-cycle assessment was applied to the phases of mineral-extraction, component-manufacturing, use and end-of-life. An important factor was a complete model simulation covering three different driving cycles. The life-cycle assessment measured the potential environmental benefits arising from use of the lightweight aluminium high-pressure die-casting cross-beam suspension. Important factors were the large amount of energy required by electrolytic production of the primary aluminium ingot and by the steel foundry. The analysis assumed a life-time mileage of 350000km.

The careful re-design of the aluminium component was essential with regard to matching the high-pressure die-cast manufacturing process and maximizing weight reduction. The primary aluminium suspension cross-beam and the steel-sheet suspension cross-beam had quite different shapes and weights but satisfied the same safety requirements in terms of structural and fatigue resistance for the same vehicle.

Assuming a diesel calorific value of 9.86kWh/l over the chosen service-life mileage of 350000 km, the results were presented in terms of the reduction in fuel consumption and emissions over a distance of 100km. Consideration of the lightened vehicle for the above driving cycles revealed, reductions in both the average fuel consumption and the exhaust emissions of 0.14%/100km. The reduction in emissions was modest because it involved the weight-saving effect of only one component, with about 15kg being subtracted from a total vehicle weight of 2350kg. Waste components at the end of the life cycle were supposed to be 95% recycled and 5% disposed in landfill. The steel component had the higher overall environmental impact.

Panel-type components have seen the main use of aluminium alloys, but remain difficult to hot-form. Two heat-treatable aluminium alloys, medium-strength 6082 and high-strength 7021 have been subjected[119] to hot stamping in order to form and quench parts simultaneously. It was possible to hot-stamp a full-scale structural part within about 30s. Differential scanning calorimetry was used in order to optimize the soaking conditions, and artificial aging restored the T6-temper strength of the as-received sheet.

The highest-strength aluminium alloys require high-temperature (120 to 200C) so-called baking so as to form a high number-density of nanoparticles via solid-state precipitation. It has been found[120] that controlled room-temperature cyclic deformation is sufficient to inject vacancies continuously into the material and thus mediate the dynamic precipitation of a distribution of very fine (1 to 2nm) solute clusters. This imparted a better strength and elongation, as compared with traditional heat-treatments, in spite of a much shorter processing time. The microstructures which were formed were much more uniform than those produced by traditional heat-treatments and did not exhibit precipitate-free zones; thus rendering them potentially more resistant to damage.

It has been shown[121] that the nanostructuring of aluminium alloys can improve the fuel-economy of vehicles by reducing the weight of the electrical wiring. High-shear non-isothermal deformation affects precipitation, thereby increasing the electrical conductivity of Al-Mg-Si alloys, while also increasing the strength by over 50%.

Microstructural analysis of alloys, before and after high-shear deformation, showed how high-shear deformation changes the grain size and the precipitate-size distribution.

Aluminium alloys which are used for lightweighting ideally require a high (~10%) elongation and a moderate (130 to 200MPa) yield strength. A range of Al-Zn-Mg alloys has been developed[122] which meets those requirements. The alloys were cast, and tested following F or T4 tempering. When in the F temper, the alloys exhibited yield strengths and elongations which ranged 130 to 190MPa and from 8 to 11%, respectively. When in the T4 temper, the yield strengths and elongations ranged from 150 to 205MPa and from 10 to 14%, respectively, depending upon the composition.

Finite-element analysis has been used[123] to design lightweight aluminium-alloy vehicle bodies. Simulations showed that, when compared with the results of bending/torsional stiffness tests, the relative errors in bending stiffness and torsional stiffness were -2.45% and -3.59%, respectively; figures which were taken to confirm the correctness of the finite-element model. The concept of a generalized structural stiffness was introduced in order to evaluate the force-transfer performance of the vehicle body. Following optimization, the bending stiffness and torsional stiffness increased by 5.59% and 1.99%, respectively, while the body weight was increased by just 0.19kg.

Sheets of 6000 and 7000 aluminium alloy are of interest for lightweighting because of their high specific strengths and low densities. In order to characterize their potential utility, a roof-beam thickness was numerically optimized[124] so as to exhibit the same flexural behavior and energy-absorption in three-point bending. The AA7075-T6, AA6111-T6 and AA6111-T4 high-strength aluminium alloys, as well as DP1000, USIBOR1500 and FORTIFORM1050 high-strength steels, were compared in a numerical study.

Low-strength aluminium-magnesium (5000 series) and aluminium-magnesium-silicon (6000 series) alloys have been used for vehicle bodies, whereas high-strength aluminium-zinc-magnesium-copper (7000 series) alloys – an obvious replacement for steel structural parts - cannot be widely used, due to poor formability. Recent studies[125] of the formability of the 7000 series aluminum alloys have however yielded positive results with regard to cold-forming, warm-forming and hot forming; especially the hot-forming quench process.

A hybrid composite was prepared[126] which consisted of an aluminium shell, and a magnesium-20wt% glass microballoon composite core, by combining powder-metallurgy and disintegrated melt deposition techniques. There was a 13% reduction in density, with respect to aluminium. There was a reasonable interfacial integrity between the aluminium shell and the Mg plus glass-microballoon core. The interface region had a Vickers hardness of 109, as compared to the hardness (68) of the aluminium shell region and the hardness (174) of the magnesium-20wt% glass core region. When compared with as-cast aluminium, the yield strength and ultimate compressive strength of the as-cast composite were increased by about 65.4% and about 60%, respectively. The energy absorption under compressive loading for the composite was about 26% higher than that for pure aluminium.

A finite element analysis was performed[127] of the chassis in seeking light-weighting possibilities in key aspects of vehicle design. These included the longitudinal tubes for body torsioning and the front crash structure. The combination of analytical results was used to develop a virtual model using finite-element methods and the model was updated on their basis. It was found that changing 6061-T651 aluminium alloy to EN-MB10020 magnesium alloy allowed the vehicle mass to be reduced by some 110kg; thus leading to a 10% improvement in fuel economy. The results implied that the current beam design, made from magnesium, would however perform badly during a crash because the force which was required to buckle the beam was the lowest (95.2 to 134kN). Steel required the largest force for buckling (317 to 540kN).

Magnesium

Magnesium alloys have been used in military applications since the 1940s for the lightweighting of materiel, in spite of corrosion and flammability problems and poor mechanical behavior, and so the use of magnesium alloys for the lightweighting of commercial vehicles requires some caution. A magnesium alloy which exhibits an extrudability and other properties that are similar to those of 6000-series aluminium alloys has nevertheless been developed[128]. The laser-heating of magnesium alloys prior to self-piercing riveting also permitted the reliable joining of magnesium components or the joining of magnesium to dissimilar metals.

The self-piercing riveting method is environmentally friendly and generates no fumes, no sparks and little noise. It can join similar and dissimilar materials. It requires no pre-drilled or punched holes and no alignment. No surface pre-treatment is required. It can form joint in the presence of lubricants and adhesives. It has a low energy requirement. More than 200000 joints can be made before tool before replacement is required. It is easy to automate and it requires a cycle-time of only 1 to 4s. The joints are water-tight and, being a cold process, there are no heat effects on the substrate materials. It ensures high static and fatigue joint strengths.

Table 20. Low-cycle fatigue parameters of GW103K alloy

Material Condition	Parameter	Value
as-extruded	cyclic yield strength	215MPa
T5	cyclic yield strength	305MPa
T6	cyclic yield strength	350MPa
as-extruded	cyclic strain-hardening exponent	0.19
T5	cyclic strain-hardening exponent	0.16
T6	cyclic strain-hardening exponent	0.20
as-extruded	cyclic strength coefficient	734MPa
T5	cyclic strength coefficient	897MPa
T6	cyclic strength coefficient	1230MPa
as-extruded	fatigue strength coefficient	518MPa
T5	fatigue strength coefficient	811MPa
T6	fatigue strength coefficient	1054MPa
as-extruded	fatigue strength exponent	-0.11
T5	fatigue strength exponent	-0.14
T6	fatigue strength exponent	-0.19
as-extruded	fatigue ductility coefficient	0.05
T5	fatigue ductility coefficient	0.52
T6	fatigue ductility coefficient	0.41
as-extruded	fatigue ductility exponent	-0.44
T5	fatigue ductility exponent	-0.87
T6	fatigue ductility exponent	-0.89

Among the disadvantages however are the facts that two-side access is usually required, although a single-side access self-piercing riveting process exists, and that a button is left on one side. Additional costs and weight also arise from the rivets. There is the possibility of galvanic corrosion occurring between steel rivets and an aluminium-alloy substrate unless the rivet has a protective coating. Finally it is not suitable for use on brittle materials such as press-hardened steel and a relatively high rivet-insertion force is required.

Because magnesium is anodic with respect to all of the other structural metals, this is a potential barrier to the use of magnesium components in vehicle-lightweighting. The effectiveness of various corrosion-mitigation strategies with regard to joined-plate assemblies has been considered[129]. Experiments revealed the interaction of magnesium, steel and aluminium, together with the effects of pre-treatments and joining methods.

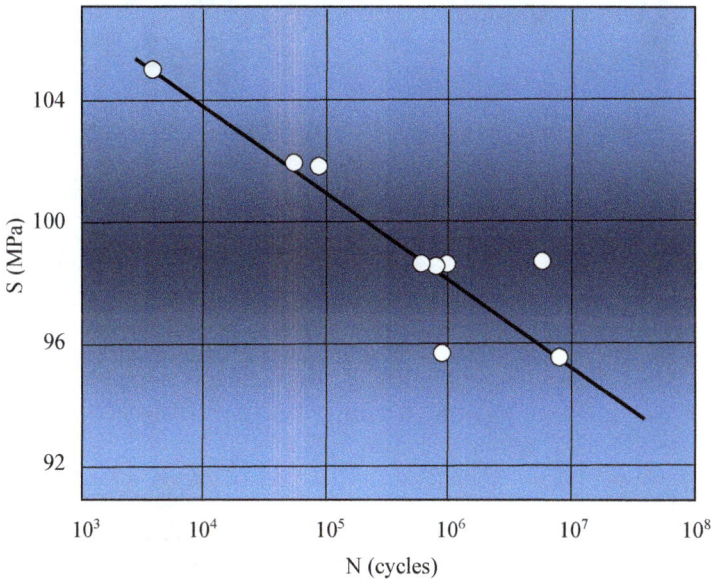

Figure 8. Fatigue curve for die-cast AZ91D alloy

Advanced magnesium alloys are favoured for light-weighting applications mainly because of their specific strength. The mechanical behavior, workability and microstructural changes during hot-deformation of these materials has been carefully studied[130]. There are advantages in using processing maps, based upon dynamic materials modelling, in order to select the optimum hot-working parameters for industrial bulk-forming operations. The grain-size evolution during deformation conditions which are suitable for dynamic crystallization follows the predictions of models based upon interface formation and migration.

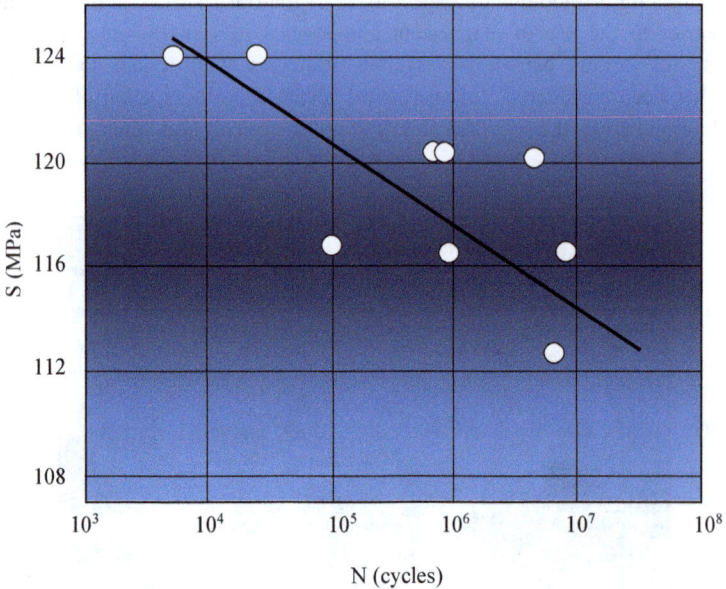

Figure 9. Fatigue curve for die-cast AZ91D alloy containing 1%Ce

Magnesium alloys have a high strength-to-weight ratio, and exhibit dimensional stability together with good machinability and recyclability. On the other hand, the hexagonal close-packed crystal structure provides only a limited number of slip systems and the alloys develop deformation textures which are associated with a marked mechanical anisotropy and a tension-compression yield asymmetry. The latter was caused by the occurrence of twinning in compression and de-twinning in tension when loaded along the extrusion or rolling direction. If vehicle components were to be subjected to dynamic loading, such asymmetries could have an unfavorable effect upon the behavior of the material. This problem could be solved by weakening the texture by adding rare-earth elements. The available fatigue properties, including stress-controlled fatigue strength, strain-controlled cyclic deformation and fatigue-crack propagation behavior have been summarized[131]. The microstructural changes and crystallographic texture-weakening in rare-earth containing cast, extruded and heat-treated material were also described[132].

Studies have shown that rare earths can be used to modify the mechanical properties of magnesium, due to their quite high solubility at the eutectic temperature and due to the formation of precipitates such as $Mg_5RE(Gd,Y)$. The as-cast alloy comprises mainly continuous rosette-shaped equiaxed dendrites and co-existing partially interdendritic eutectic. The main intermetallic compounds are $Mg5RE$, $Mg_{41}RE_5$ and $Mg_{24}RE_5$. The average equiaxed interdendritic arm spacing in a typical casting was about 30µm for Mg-15Gd-5Y-0.5Zr alloy; smaller than that in AZ31 and AM30. This was because the rare earths and zirconium. Grain refinement has also resulted from adding gadolinium, yttrium, cerium and neodymium.

Wrought magnesium alloys have better mechanical properties than those of the cast equivalent due to hot-deformation during extrusion, rolling and forging. During hot-extrusion recrystallized grains form on the grain boundaries, possibly due to an accumulation of dislocations.

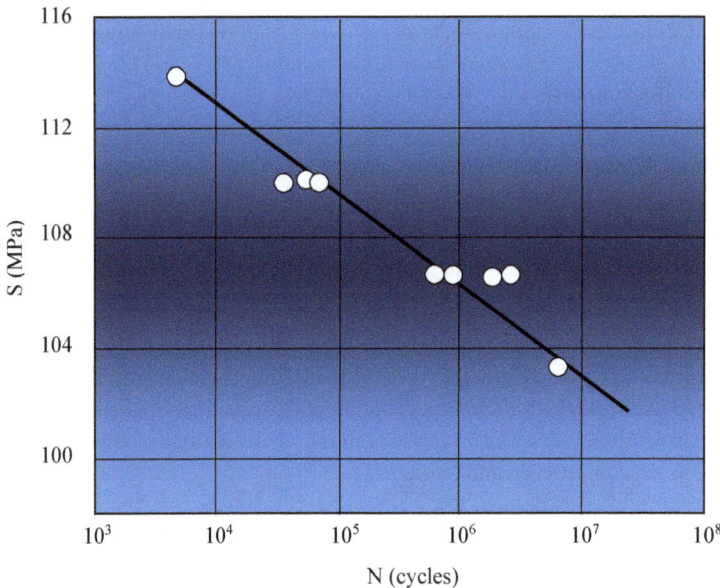

Figure 10. Fatigue curve for die-cast AZ91D alloy containing 2%Ce

The magnesium-rare-earth alloys also exhibit a rapid response to precipitation-hardening, due possibly to the presence of a high volume fraction of metastable precipitates. The good mechanical properties of the alloys are attributed mainly to metastable and stable precipitates which preserve their stability at relatively high temperatures.

The fatigue behavior of magnesium alloys is of relevance to lightweighting and is best summarized by traditional stress-controlled fatigue data in the form of so-called S–N curves, where S is the stress amplitude and N is the number of cycles to failure. Figures 8 to 10 are high-frequency S-N plots for AZ91D alloy containing various amounts of cerium. The addition of 1 or 2wt%Ce clearly results in improvements in fatigue strength; 20.3 and 9.1%, respectively. Similar improvements have been found upon adding neodymium, with or without cerium. The main factors which govern the high-cycle fatigue properties of die-cast AZ91D appear to be the grain size and the grain-size distribution. The grain size tends to be greatly refined due to the rare-earth additions. This improves the fatigue properties because the number of grain boundaries increases with decreasing grain size, and that increases the resistance to dislocation movement during plastic deformation. Cerium additions also decrease the size and amount of porosity or inclusions.

Strain-controlled low-cycle fatigue can occur when the repeated stresses are heat-produced and high stress levels occur over a low number of cycles. Most studies have involved the strain-controlled low-cycle behavior of Mg–Gd–Y alloys. The GW83 alloy (table 20) exhibited almost symmetrical stress–strain hysteresis loops, marginal cyclic hardening and an almost zero mean stress.

The magnesium-rare-earth alloys have longer fatigue lives when compared with those of wrought alloys which do not contain rare earths. The Mg-10Gd-2Y-0.5Zr (GW102K) alloy has a better fatigue resistance than that of AZ31, for a given strain-amplitude.

The GW103K alloy has an increasing fatigue life with decreasing strain amplitude under any conditions. In polycrystalline hexagonal close-packed magnesium alloys mechanical properties such as the cyclic deformation behavior are directly related to the presence of crystallographic textures which dominate the orientation of slip and twinning planes and directions relative to an externally applied stress. This naturally affects the yield asymmetry between compression and tension.

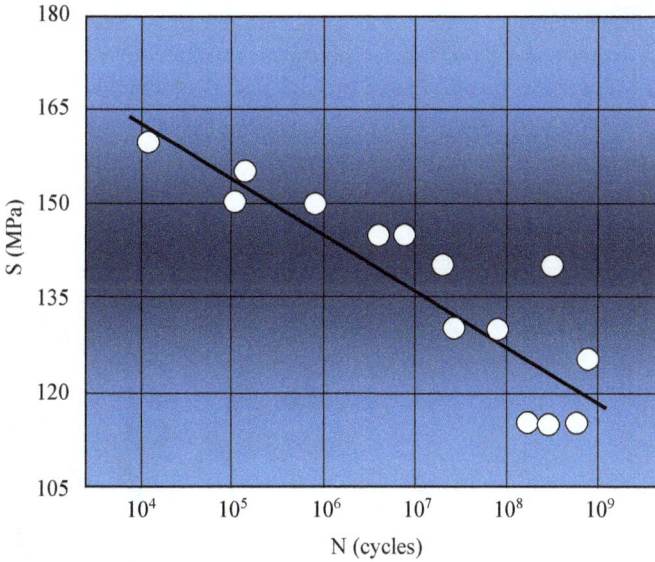

Figure 11. Fatigue curve for extruded GW123K (Mg-12Gd-3Y-0.5Zr)

Twinning is a key deformation mechanism in magnesium alloys and a combination of twinning and crystallographic textures in wrought magnesium alloys is responsible for a tension–compression yield asymmetry that is common to nearly all magnesium alloys. It is also suggested that the yield asymmetry is sensitive to textures. On the other hand randomly textured cast samples, vacuum die-cast samples and semi-solid processed samples exhibit almost no yield asymmetry. An important method for modifying the yield asymmetry and twinning is to alter the texture. Because rare earths weaken the texture, this could lead to an absence of tension-compression yield asymmetry.

Zener-pinning of grain boundaries, caused by rare earth elements which are present in solid solution or as particles might play a role in restricting the preferential growth of some grains, enabling the growth of diverse grain orientations and leading to a weakening of the texture. Such pinning would increase in influence with decreasing particle-size. A smaller grain size, and a stronger Zener effect, could lead to a relatively weak texture and almost-symmetrical hysteresis loops.

Materials Research Forum LLC
https://doi.org/10.21741/9781644902134

A marked change in deformation-mode, from mainly twinning in rare-earth free AM30 alloy to mainly dislocation slip in rare-earth containing GW103K can be attributed to grain-refinement, large uniformly-distributed precipitates and texture-weakening. Their combined effect led to a sharp reduction in hysteresis loop asymmetry, an increase in stress amplitude, a decrease in plastic strain amplitude and an increase in fatigue-life.

Table 21. Properties of common magnesium alloys

Alloy	ρ (g/cm^3)	UTS (MPa)	YS (MPa)	e (%)	E (GPa)	Specific Strength (MPacm^3/g)
AZ91	1.82	280	160	8	45	154
AM60	1.79	270	140	15	45	151
AZ31	1.76	300	180	12	44	170
AZ80	1.8	330	210	10	45	183
ZK60	1.81	340	280	10	43	188
KBM10	1.79	360	250	23	45	201
ZW61	1.81	420	250	16	45	232

Given that fatigue-fracture in metals proceeds via the stages of crack-initiation, crack-propagation and final rapid failure, some 10% of fatigue life has usually elapsed before a crack nucleates, and this normally occurs at a free surface. The flow lines or tearing ridges which are indicative of the fatigue crack growth direction have been be seen in GW103K in the as-extruded and T5 state. The crack propagation region was characterized by typical fatigue striations perpendicular to the crack-propagation direction. A mixture of cleavage-like features and tearing ridges appeared near to the initiation-fracture site in T6 samples fatigued at a strain amplitude of 0.4%. The spacing of fatigue striations became steadily greater with increasing distance from the initiation site, with each striation presumed to be associated with a single loading cycle. Assuming that fatigue striations are due to repeated plastic blunting-sharpening events due to the slip of dislocations in the plastic zone ahead of the crack tip. The formation of striations in the close-packed hexagonal magnesium alloy was expected to be related to both dislocation slip and twinning in the plastic zone, and to be similar to those occurring in rare-earth free magnesium alloys.

Studies of gadolinium- and yttrium-containing alloys such as extruded Mg-12Gd-3Y-0.5Zr (figure 11) indicate that it exhibits a continuously decreasing S–N curve with no horizontal asymptote and possesses a higher fatigue strength than that of AZ31. The addition of the rare earths caused the elimination of the usual tension–compression yield asymmetry, found in conventional wrought magnesium alloys, by texture randomization and the avoidance of deformation twins by basal-slip activation.

Attention has been paid to the use of sheet magnesium (table 21) to manufacture panels and structural components, given that its density is some 40% lower than that of aluminium and 75% lower than that of steel. Investigations have also been made[133] of changes in the microstructure during the thermomechanical processing of twin-roll cast AZ31B alloy sheets (table 22), and of the mechanical properties under superplastic conditions. Rate-dependent crystal plasticity models have been coupled with the Taylor polycrystal model in order to capture the overall behavior of the material.

Table 22. Tensile properties of AZ31B sheet

Orientation wrt Rolling Direction (°)	YS (MPa)	UTS (MPa)	e (%)
0	207	291	21.5
45	242	299	24.8
90	265	310	24.5

The cross-car beam assembly of a vehicle was redesigned[134] with the aim of substituting an extruded magnesium alloy for steel. Extruded and stamped parts were designed separately, due to the different processes applied to the material, and joined later. The results showed that magnesium alloy substitution met the requirements of lightweighting and collision safety. A structural evaluation was made[135] of material design characteristics such as bending stiffness, torsional stiffness and crashworthiness in order to decide whether magnesium truly provides a better alternative to the current use of aluminium in the automobile industry. It was noted that there could for instance be a problem with a rocker-beam width/thickness ratio, leading to failure in yield rather than buckling. By replacing the chosen material, 6061-T651 aluminium alloy, with EN-MB10020 magnesium alloy, the vehicle weight could be reduced by about 110kg; thus improving the fuel economy by 10%. This would require a compromise to be made in the mechanical performance however unless the design was also modified. In another case[136], the use of a magnesium die-casting as a replacement part resulted in a vehicle liftgate-assembly weight-reduction of almost 50%. The introduction of a cast AZ91 magnesium-alloy wheel has been most recently considered[137]. Numerical simulation of the casting process could shorten the manufacturing process and reduce the cost as well as the incidence of casting defects. Finite-element methods predicted the mould-filling and solidification behavior and, in order to minimize casting defects, an improved riser-system was designed. Significant reductions in vehicle weight might be made by using magnesium alloy to construct the wheels. A new design of wheel has been described[138], together with a finite-element model for bending and radial tests. Magnesium alloys, aluminium alloys and steels were all considered for its construction and the associated stresses and strains were predicted. Evaluation of the model in bending and radial tests predicted a better behavior for a magnesium-alloy wheel.

Lightweight metal castings can result in reduced manufacturing costs of entire components and assemblies as compared to those of stamped and wrought fabricated assemblies due to the complex shapes that can be manufactured using various processes and the cost benefits that accrue from parts integration: fewer parts and fewer joints. The main technical challenge to the casting of complex components is the maintenance of reliably adequate properties for the component while minimizing expense. A life-cycle assessment was made[139] of an engine block that could be used for light-weighting. At each stage of production the environmental effect of using more sustainable methods was considered. This included the use of a low-impact cover gas, for protecting the magnesium melt from oxidation, instead of high-impact sulphur hexafluoride.

One project has been aimed entirely at enabling the use of magnesium in automotive lightweighting, with the objective of developing those technologies required to introduce high-volume, cast structural magnesium components that could function in harsh environments such as an automobile chassis. This involved simulation models, joining techniques, corrosion-resistance methods, fracture mechanics and non-destructive evaluation. As a practical application, a front engine cradle was selected because of its difficult highly-loaded under-body location. The magnesium engine cradle casting was intended to mesh with re-design, load analysis, process simulation and casting trials. The cradle went into volume production and resulted in a 34% weight-saving over the existing aluminium version.

A redesigned rear cradle was optimized[140] by using metamodel techniques together with large-scale finite-element simulation. The cradle was redesigned so as to accommodate an electric drive system in order to convert the vehicle into a hybrid. The design-optimization used metamodels to represent the finite-element responses of the cradle under multiple loading conditions. The metamodels reduced design-time and the number of required simulations. The result of the optimization was a 12% weight reduction combined with a higher confidence level with regard to safety.

The objective was to determine the feasibility of reducing the mass of power-train components by at least 15% by replacing cast aluminium with cast magnesium. The project included the engine design, alloy selection for each component, and tooling design and casting of all of the magnesium components: the sand-cast cylinder block, the high-pressure die-cast structural oil pan and front engine cover, and the thixomoulded rear seal carrier. The results well exceeded the mass reduction target with respect to the aluminium version.

The high-integrity magnesium automotive castings project investigated four casting processes: low-pressure permanent mould and squeeze-casting processes which were modified current aluminium-casting methods, plus and two new processes. The objective was to develop metal-casting processes which could cost-effectively manufacture components with high ductility, high strength, low porosity, oxide-free, inclusion-free from cast magnesium. These automotive chassis components were expected to have geometries and properties were not achievable by using existing high-pressure die-casting processes. This involved the concurrent development of key materials and manufacturing techniques which would permit the design and use of magnesium-based automotive-body front-end structures of greatly reduced weight but equivalent performance and cost when compared with sheet-steel structures. This was the first project to focus on an entire sub-assembly; a front end consisting of sheet-metal, extruded and cast components, joined using various processes to magnesium and steel or aluminium. Given that the front end plays a crash-energy containment role, it was also important to evaluate the crashworthiness of the cast or wrought components. The casting processes which were evaluated were vacuum die-casting, squeeze-casting, low-pressure die-casting and thixomoulding. The application of magnesium-sheet in vehicle structures is hindered by the limited formability of magnesium sheet and the inherent cost of producing the sheet. Warm-forming can be used to improve greatly the formability of magnesium sheet. A warm-forming cell was built in order to demonstrate the efficient forming of magnesium sheet. Another example was direct-chill cast AZ31. The intended application was deep-drawn panels for use as door-inners.

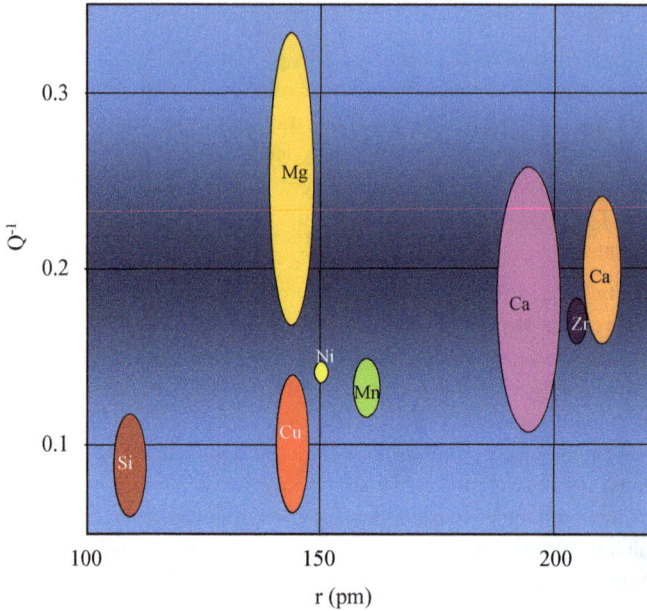

*Figure 12. Relationship between damping capacity and
radius of alloying elements for binary magnesium alloys*

The cost of magnesium sheet is governed by the high costs of rolling an ingot into sheet form, and this in turn is a direct result of its hexagonal close-packed structure because it requires the sheet to be rolled in small increments; often with intervening annealing steps between the rolling passes. The use of continuous casting may markedly reduce the cost because, by casting the material directly as sheet, continuous casting results in a higher production rate, lower capital investment and much less energy when compared with conventional direct-chill ingot-casting. The new warm-forming system was used as a benchmark for the evaluation of new magnesium sheet materials.

Magnesium alloys also have potential value in railway rolling stock for the same reasons. High-strength magnesium-alloy components are particularly important to lightweighting efforts in the railroad industry, with a view to energy saving and reductions in friction/wear, vibration and fatigue damage. A magnesium-alloy train has a theoretical

8.6 to 12.6% weight-reduction potential in the equal-strength and equal-stiffness condition where low-speed trains exhibit greater energy-savings than do high-speed trains[141]. Magnesium alloys have a strength advantage over other alloys and their good damping characteristics (figure 12) naturally decrease vibration.

The so-called high-solubility alloys, with elements such as manganese, zirconium, calcium and yttrium lead to a greater damping capacity than so-called low-solubility alloys (silicon, copper, nickel). Those solute atoms, such as yttrium, which have a large size-misfit with respect to the magnesium atoms and have a larger solid solubility can generate more pinning-points for dislocation motion, leading to a better damping performance. Annealing and aging can further improve the damping properties.

Although it will not be considered in detail here, efforts made in the lightweighting of railway stock offer valuable clues to achieving the same thing in the case of automobiles. It can be noted in passing that the lightweighting of shipping containers is a neglected strategy for saving energy and reducing greenhouse-gas emissions. The life-cycle fuel savings and environmental aspects of the lightweighting of a 12.2m container have however been examined[142]. Use-phase stages for conventional and lightweighted (steel-reduction, aluminium-substitution or high-tensile steel substitution) containers were compared. A lifetime reduction of 21% in the fuel required to transport a container, and a 1.4% reduction in the total fuel required to move a vehicle, cargo and container, were predicted. A 10% reduction in the weight of the system was expected to result in fuel reductions ranging from 2 to 8.4%. At the global scale, container-lightweighting might reduce the energy demand by 3.6EJ and greenhouse-gas emissions by 300,000,000tonnes$_{CO2}$[eq] over a 15-year lifetime. A statistical analysis[143] of two heavy goods vehicle fleets showed that reducing the empty weight of trailers by 30% can cause reductions of up to 18% in the mass energy performance index for double-deck trailers.

The two obvious strategies in both cases is either to reduce the absolute size and/or the density of the material used. Study of the first approach[144] showed that, if the outer diameter of a hollow axle were to be increased by 30% with respect to that a solid axle, a weight reduction of 56% could be achieved by using a hollow steel axle. Further weight-savings could be achieved by changing the material which used for the hollow axle. A carbon-fiber composite might lead to a 64% weight-saving when compared with a hollow steel axle and an 84% weight-saving when compared to a solid steel axle. Axles in particular have been investigated[145] with the aim of reducing the unsprung weight of a rail vehicle. Such a reduction helps to reduce track damage, and would presumably also reduce road damage, although this is not usually a major concern in the case of vehicles. The design of a composite railway axle was part of a parametric study of a tube axle that indicated the greatest potential for weight reduction. An existing hollow steel axle was used as a benchmark for weight, strength and stiffness, and the estimated weight of a composite axle was 50kg; yielding a weight-reduction of 75%. Five components: a cantilevered seat-bracket, a luggage-rack module, and intermediate end-structure, a body side-structure and a roof-structure have been identified[146] as being the most suitable for demonstrating the benefit of composite-replacement of rolling-stock paraphernalia. It has been estimated that composite re-design of these items would result in weight-savings of 57% in the case of the intermediate end-structures, 47% in the case of the body side-structures and 51% in the case of the roof structures. One can envisage seeking to make similar weight-savings in the road-vehicle case.

Advanced forming techniques are being investigated that offer the potential to increase the forming rate used for aluminium alloys in order to meet the high-volume part production required for the automotive industry and to improve the resultant mechanical properties of components formed at the requisite higher temperatures to achieve such rates. As noted elsewhere, warm-forming techniques are being developed for direct-chill and continuous cast magnesium sheet in order to improve formability and lower the cost of production.

Another project involves the electromagnetic forming of aluminium sheet into the required shape. In this process a transient high-magnitude electrical pulse is directed into a specially designed forming coil via a low-inductance electrical circuit. During the current pulse the coil is surrounded by a strong transient magnetic field. As in the case of the well-known 'jumping-ring' student experiment, the transient magnetic field induces a current in the nearby conductive workpiece which flows in a direction opposite to that of the current in the coil. The coil and workpiece then have parallel currents running through the two conductors, which thus repel one another. The repulsion force can be equivalent to surface pressures of the order of tens of thousands of pounds per square inch, and thin sheets of material can be accelerated to a high velocity within less than a millisecond. The high workpiece velocity and resultant strain-rate enhances the formability of materials such as aluminium and the dynamics of contact with the forming die help to reduce spring-back. This old technique has usually been applied to tubes and the forming of sheet material is much more complicated. It is expected that electromagnetic forming is best hybridized with conventional sheet-metal stamping.

Another proposed[147] lightweight axle assembly would consist of a carbon-fiber reinforced polymer composite tube with steel stub axles at each end. The hybrid metallic-composite railway axle would then be surrounded by various coaxial skins. The optimized design comprised a composite tube with an outer diameter of 225mm, a composite tube thickness of 7mm, a steel stub-axle extension thickness of 10mm and a bond overlap length of 100mm. This axle had a weight of 200kg; yielding a weight-reduction of 50%.

Traditional freight wagons have I-beam sections as the main load-bearing structures, and changes in their design are less obviously transferrable to vehicles. There is nevertheless considerable scope for the design of superior beam sections[148]. A novel railcar has been designed[149] in which a self-powered very light vehicle has two bogies; each containing a complete diesel-electric series-hybrid drive system. The entire vehicle was lightweighted so as to reach a target weight of less than 18 tons. The lightweighting led to a reduction in the propulsion requirements and reduced the infrastructure, installation and maintenance costs. The use of higher-efficiency drive systems and on-board energy systems also reduced the CO_2 emissions.

In order to design the lightest-weight components possible, lifetime prediction can be improved with the aid of crack-growth theory. That is, it is assumed that all metal components have a certain defect content. The defect size can be measured by using ultrasonic and other techniques. Processing can also be improved so as to minimize the defect size with the aid of vacuum-arc re-melting, given that inclusions result from the addition of elements which increase machinability and the vacuum-arc process minimizes inclusion formation. Life-prediction can be based upon the material properties, residual stresses and initial flaw sizes. The fracture mechanics technique has been applied to a standard hypoid axle gear design so as to determine the lifetime that can be achieved by using cleaner steels, single-peening and dual-peening. The initial size of the ring gear is a determinant of the weight of the entire axle assembly in that, if the size can be reduced, the other components will also exhibit a consequent weight reduction. Such improvements thus promote appreciable fuel and emissions reductions.

The weight of a rear axle-shaft has been reduced[150] by using a hollow-shaft geometry and intensive quenching. Load-mapping was used to set up a finite-element model which could then be used to evaluate various lightweighting designs and techniques. Weight-savings of some 26% could be achieved by replacing the usual solid axle-shaft with a hollow shaft. Intensive-quenching was better than oil-quenching with regard to residual stress levels and strength, in that it allowed the removal of 3% of the shaft weight. Compressive residual stresses which were created at the surface during intensive quenching acted so as to slow crack nucleation and increase the fatigue life.

The casting of aluminium has many attractive attributes as a process including the ability to produce complex shapes that might replace several simpler parts. The wall thickness of castings is dictated not only by performance requirements for the part in use, but more often by the processing limitations. The process constraint can limit the mass savings that can be realized in an aluminium casting. Super-vacuum die-casting is a method that removes the air from the die cavity immediately before filling.

The resultant properties include excellent levels of yield strength, ultimate tensile strength and elongation. These changes can lead to parts having a lower mass than those of conventionally die-cast parts. The use of super-vacuum die-casting technology involves the modification of dies and process settings, vacuum systems, valves and control systems. Because of such modifications, the method may be prohibitively expensive. An analytical model has been developed in order to explore the cost difference between super-vacuum die-casting and conventional die-casting.

It was found that the economic advantages of super-vacuum die-casting were greatest when it permitted a part to be designed so as to have a reduced mass. The highest additional costs were due to die-cast tooling for the vacuum technology. It was concluded that super-vacuum die-casting is more economically advantageous for larger parts and high-volume production. In the case of a small part produced in small quantities, the cost break-even point can require a material saving of 40% to be made. In the case of a large size part which is to produced in high volumes, break-even point requires a mass-saving of only 10% to be made.

During conventional high-pressure die-casting a significant amount of air is entrapped in the die cavity and that air produces air-pockets within a cast part that impair the mechanical properties of the part; especially after quenching. The entrapped air can also contain hydrogen gas and residual die-lubricant. The pressure within the cavity during conventional die-casting is about 1600mbar, and is much lower when a vacuum is created. In a typical vacuum-assisted die-casting process, the pressure is 100 to 200mbar within the die cavity.

As noted before, the presence of gas in a cast part affects the mechanical properties and leads to the design of thicker parts in order to offset any impairment of the mechanical properties. The parts are then unavoidably heavier. Studies involving pressure levels of between 180 and 280mbar revealed a marked reduction in the size and number of gas-pores and improved mechanical properties. Vacuum-cast specimens also suffered had no visible blistering, unlike castings which were made without vacuum-assistance.

The pore-size was also reduced, and they were evenly distributed in vacuum-cast specimens but were larger in size, and clumped, in the other specimens. There was no great difference in the hardness of the two specimens; but the tensile strength was 10% greater in the case of vacuum-assisted castings. Super-vacuum die-casting of Mg-Al-Mn alloy was compared with conventional high-pressure die-casting, showing that the super-vacuum die-castings had significantly better mechanical properties, with the bending fatigue strength being improved by 64% and the corrosion resistance being improved by 39 to 62%.

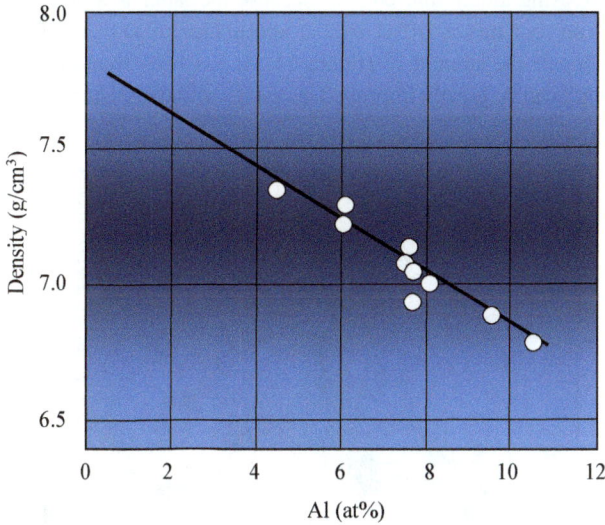

Figure 13. Density as a function of concentration in Fe-C alloys

The elimination of air from the die cavity allows castings to be thin-walled and reduces the thickness by up to 40% and the weight by up to 20%. These results can be achieved when super-vacuum die-casting is used with novel materials. In the case of conventional high-pressure die-casting no heat treatment is usually carried out as part of the process. When vacuum is added to high-pressure die-casting heat treatment would add to the cost. Optimum mechanical properties can however be achieved only when cast parts are heat treated, and heat treatment is possible only in the presence of minimal porosity. Super-vacuum die-casting also has a positive downstream effect upon scrap rates because casting defects such as porosity and shrinkage are largely avoided.

Steel

The idea of reducing the density of steel is far from being a new one[151] and the obvious strategy is to add aluminium, even though that element is known to have deleterious effects upon conventional castings. The addition of aluminium to plain carbon steels reduces the density (figure 13), but also the Young's modulus, with the latter decrement being about twice as great per 1% of aluminium added. The yield strength initially increases with increasing aluminium content because the latter is a good solid-solution strengthener for iron, but this is offset by the appearance of D0$_3$ and B2 superlattice structures which impair the strength[152] (figure 14). It has been found however that the ductility and strength can be greatly improved by means of thermomechanical treatment.

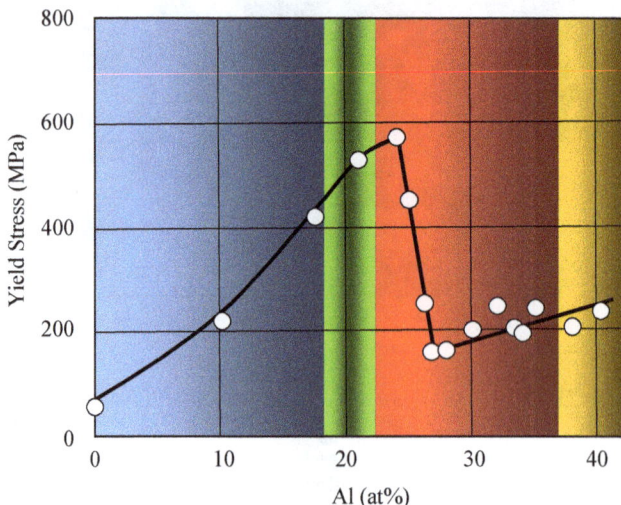

Figure 14. Yield stress as a function of concentration and structure in Fe-C alloys
Blue: A2, green: A2+D0$_3$, red: D0$_3$, yellow: B2

Some studies have suggested that steel is the only material which can satisfy the conflicting demands on the properties required for vehicular applications. High-strength steels are commonly used to reduce mass and improve structural performance. Between 2010 and 2013, the use of advanced high-strength steels increased by 7.1kg per vehicle[153]. The introduction of lightweighting materials into vehicles is not however a simple matter of removal and replacement. Factors such as manufacturing feasibility have to be considered. Modification of stamping, tooling and joining methods permit an easier transition to the use of new grades of steel.

A lightweight design for steel vehicle-bodies was proposed some time ago[154] which offered appreciable weight-savings, could be constructed by using existing technology and was essentially cost-neutral. One of the most effective routes to vehicle-lightweighting involves the hot-forming of ultra high-strength 22MnB5 boron-steel. Dynamic recovery is the main softening mechanism of the steel at high temperatures when in the austenitic state. Uniaxial tensile-testing of the boron-steel was carried out[155] at high temperatures. True stress-strain curves, and the relationship between the work-hardening rate and the flow stress were obtained under various deformation conditions. It was found that the work-hardening rate decreased linearly with increasing flow stress. A flow-stress model was derived on the basis of the Kock model and the derivative of the dislocation density with respect to the true strain, expressed by the peak stress and initial yield stress, was also deduced with regard to the dynamic recovery effect. A dynamic recovery efficiency factor was defined as being the ratio of the dynamic recovery effect and the dislocation accumulation effect. High-manganese twinning-induced plasticity steel exhibits both a high strength and a good formability, and offers the possibility of altering the mechanical properties by modifying the strain hardening[156]. The use of twinning-induced plasticity steel can greatly aid the lightweighting of steel components, as well as reducing material use and improving the press-forming behavior. These steels are of particular importance with regard to vehicles which are equipped with hybrid or electric motors, because these are considerably heavier than conventional motors. They can also satisfy component requirements with regard to crash performance, and their high tensile strengths can cope with the cyclic loads which act on the chassis. It is thus of great interest to improve their fatigue strengths by carrying out pre-straining and surface treatments. It can however require a lot of time to optimise such treatments, thus hindering material development. A new method[157] has been developed which permits rapid and accurate determination of the fatigue strength and is based upon monitoring the changes in the stiffness of a specimen at various stress levels in order to monitor the evolution of fatigue damage. Predictions made using the new approach have been confirmed by comparison with conventional staircase results, and standardised fatigue-crack growth tests, revealing good agreement.

The choice of material may be limited in the case of certain critical components, due to strength and collision-resistance requirements. This is true of the B-pillar, which is a vertical support located between the front and rear sections of a vehicle and is critically important in ensuring structural integrity during side-on impact. Lightweighting can thus be pursued by using high-strength steels which can provide an equal or superior collision-resistance while also requiring a thinner and lighter cross-section. Such advanced can nevertheless be difficult to stamp or weld and potentially suffer from delayed cracking. An important innovation has been the development of low-carbon base compositions that are micro-alloyed with niobium. A comparison has been made[158] of the life-cycle impact of two advanced steels, proposed for the construction of the B-pillar: a press-hardened boron steel and a hydroformed combination of the molybdenum-containing dual-phase steels (DP800, DP1000).

The DP800/DP1000 B-pillar option was associated with a lower environmental impact, in that the global-warming risk of this choice over the entire life-cycle of the vehicle was 29% lower than that of the boron-steel option. The in-use phase was the main culprit, and accounted for 93% of the life-cycle global-warming effect. A weight-saving of just 4kg accounted for most of this predicted difference in the effect of the competing B-pillar choices.

The harm arising from the manufacturing step was also lower for the DP800/DP1000 design, in spite of its higher alloy content: even if the effect of hydroforming was twice that of press-hardening, the global-warming resulting from production of the DP800/DP1000 pillar design was predicted to be lower than that of the other choice.

The environmental damage which is associated with the body-structure of a vehicle can therefore be reduced by an increased use of sophisticated high-strength steels without lessening impact-resistance. The relationship between crushing force, material strength and thickness of thin-walled beams under three-point bending has been determined[159], and this informed a lightweighting strategy for B-pillar design (tables 23 and 24). The lower portion of a B-pillar is crushed and bent during a side-on impact, and this can be simulated by performing three-point bend tests. A high-strength steel of high ductility can here be used instead of a weaker conventional high-strength.

Rigid rotation of the upper portion of the B-pillar occurs during a side-on collision and is best analyzed as a problem in statics. A static force which simulates a side-on collision can be applied, with the upper portion being divided into a number of segments and the thickness of each segment being optimized. The latter approach leads to a lightweighting saving of 24%, without affecting the crashworthiness while the intrusion speed and material volume of the main part of the B-pillar are almost equal to those of the other design.

A life-cycle assessment of the cradle-to-grave effect of the changed design of the B-pillar compared the press-hardened boron steel and advanced high-strength options while assuming a weight of 13.3kg in both cases; the latter amounting to some 8% of the total body-weight (325kg) of the vehicle skeleton.

The new B-pillar was a hydroformed component which consisted of 76% DP800 and 24% DP1000 dual-phase steels, and had a weight of 9.3kg; a saving of 30%. The DP800 steel contained up to 1%Cr, plus a molybdenum content of about 0.18%, a manganese content of 2.5wt% and a total alloy content of 2 to 3%. The DP1000 steel contained up to

Materials Research Forum LLC

https://doi.org/10.21741/9781644902134

1.4%Cr, plus a molybdenum content of about 0.33%, a manganese content of 2.9wt% and a total alloy content of between 2 and 3%.

A typical boron steel contained between 1 and 1.4%Mn, 0.11 to 0.25%Cr and 0.0008 to 0.005%B; with a total alloy content of between 1.5 and 2.5%. The two B-pillars were equivalent to the extent that collision tests indicated the same maximum side-on collision intrusion at the base of the pillar. There was however a reduction in maximum intrusion of 64mm at the top of the pillar in the case of the new design, where the intrusion is a measure of the inward deformation of a component that occurs during a collision.

The effect of B-pillar re-design was predicted for a 10-year period and a total lifetime of 200000km. Leaving aside any differences in the construction difficulty of the competing B-pillar designs, the resultant in-use fuel consumption savings were estimated. The typical fuel-saving, in the absence of power-train re-sizing were 0.15l/(100km100kg) for petrol and 0.12l/(100km100kg) for diesel.

In the case of power-train re-sizing, these figures became 0.35l(100km100kg) and 0.28l/(100km100kg), respectively. A recycling-rate of 95% was assumed for steel at the end of life. Doubling the effect that resulted from the forming process was not expected to result in any increased global-warming risk over that of the original boron-steel design.

The effect of production with regard to acidification and eutrophication was higher for the new design when the forming process was increased by 50%. A cross-over occurred, in the case of photochemical ozone formation and primary energy demand, when the forming process impact was above 80% and above 90%, respectively. A change in the proportion of DP1000 in the B-pillar had a slightly greater effect than a change in DP800, with the predicted increase being less than 5% if the component was 100% DP1000.

Table 23. Life-cycle results for B-pillar re-design of petrol vehicle

Environmental Risk	Original Design	New Design
Global warming (kg$_{CO_2}$[eq])	278	196
Acidification (mol$_{H+}$[eq])	0.715	0.513
Eutrophication (kgP[eq])	0.000368	0.000257
Photochemical formation (kg)	0.273	0.197
Energy demand, total (MJ)	3915	2753
Energy demand, non-renwwable (MJ)	3760	2643

Impact-protection structures often consist of multiple thin-walled sections comprising metals and polymers; a complication being that, at typical vehicle operating temperatures of -40C to 80C, the mechanical properties of steel are essentially temperature-insensitive while those of polymers can exhibit three-fold changes in magnitude. This leads to differing impact responses in summer and winter for steel-plastic structures. The B-pillar tends to have a thin-walled steel section in order to satisfy stiffness and strength demands, plus a thin-walled plastic trim interior for occupant impact-protection[160]. Such a structure has a limited depth in the impact direction and its constituent materials have differing temperature properties. It is found that the impact response is very temperature-dependent and the consequent protective effect is different in winter and summer. Head-impact against a B-pillar under winter conditions is greater than that at room temperature because the plastic trim is the harder and the peak acceleration of the head is greater. The plastic trim is softer under summer conditions, but the residual kinetic energy additionally deforms the steel and leads to a higher peak acceleration and greater intrusion.

Table 24. Life-cycle results for B-pillar re-design of diesel vehicle

Environmental Risk	Original Design	New Design
Global warming (kg_{CO2}[eq])	238	168
Acidification (mol_{H+}[eq])	1.08	0.767
Eutrophication (kg_P[eq])	0.000255	0.000179
Photochemical formation (kg)	0.433	0.309
Energy demand, total (MJ)	3426	2410
Energy demand, non-renewable (MJ)	3289	2313

The acidification potential, photochemical ozone formation and eutrophication potential are all linked to airborne emissions, such as NO_x, SO_x and volatile organic compounds, which result from the burning of fuels. The eutrophication potential is an exception when the value of scrap is used to calculate the end-of-life credit due to steel-recycling because it does not have a negative effect. That is, the eutrophication potential of the electricity consumed when producing steel from scrap in an electric-arc furnace is greater than the eutrophication potential involved in producing blast-furnace steel.

Some of these new metals are difficult to machine and this consequently reduces tool-life as well as increasing the machining costs. The use of electricity can reduce the cutting force during orthogonal cutting and turning. The passage of an electric current can reduce the cutting force by 10% in the case of 1008 steel, at the cost of an increased temperature and of possible arcing during the initial contact; which again causes increased tool-wear at high current inputs. The phenomenon of electroplasticity can offset the problem that some replacement steels for lightweighting purposes exhibit with regard to reduced formability and greater degree of spring-back. Electrical treatment can decrease the spring-back tendency and increase the formability of sheet metal. The electric current which is applied is usually characterized in terms of its current-density, but this parameter does not always capture the subtleties of high strain-rate manufacturing processes. Other predictors of electrically-assisted processing behavior are the electrical energy and power. The former is a better predictor than current-density, but depends upon an ability to predict the processing temperature. The spring-back phenomenon, which is due to the elastic recovery of a material, is a problem which can prevent component assembly unless additional forming steps are carried out. In the case of bent bonded dissimilar materials, such as aluminium and composites, the spring-back depends upon which of the materials encounters the punch or die. This is a matter of importance due to the differences in the elastic properties of the materials.

Transformation-induced plasticity steels are promising choices for lightweighting purposes because of their enhanced formability. On the other hand, the predictive modeling of that formability is difficult because of the combination of dislocation and transformation mechanisms which operate during deformation along various strain paths. A crystal-plasticity constitutive model can be used[161] to simulate the behavior of deformation-induced $\gamma \rightarrow \alpha'$ transformations, and such a model can be introduced into the crystal-plasticity framework so as to describe fully the behavior of transformation-induced plasticity steels. Simulations could be compared with experimental data on duplex stainless steel, and the calibrated model could then be used to predict the forming-limit diagram by applying the Marciniak-Kuczynski approach. Transformation-induced plasticity steels exhibit favorable combinations of strength and ductility but the zinc-coated material suffers from low-energy fracture and liquid-metal embrittlement and this limits their use for lightweighting. Resistance spot-welding of zinc-coated steel, with an ultimate tensile strength of at least 1180MPa, produced welds that were associated with various degrees of surface cracking. The static strength and fatigue life under tensile loading were determined[162] in a cross-tension orientation. When compared with crack-free welds, no appreciable reduction in cross-tension strength, absorbed energy or fatigue life was observed when testing samples which contained cracks of less than 325μm in length. The performance of spot welds in the presence of cracks which were longer than 500μm in length was significantly impaired.

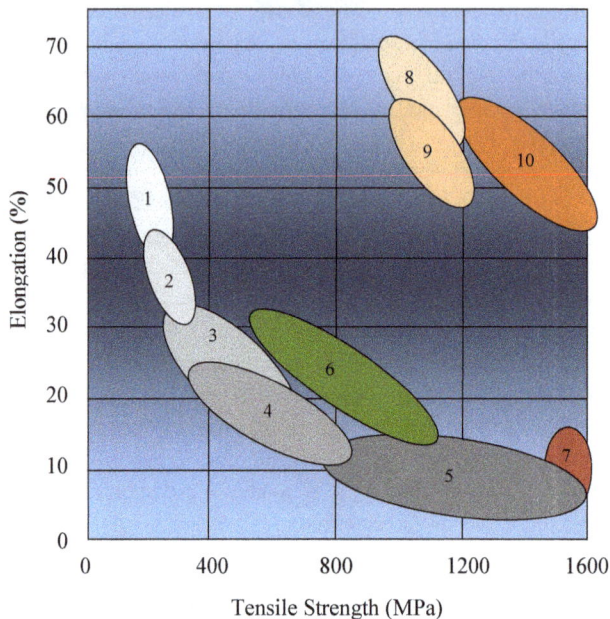

Figure 15. Elongations and tensile strengths of various steels, 1: interstitial-free, 2: mild steel, 3: high-strength low alloy steel, 4: DP, CP, 5: martensitic, 6: transformation-induced plasticity, 7: press-hardening steel, 8: L-IP. 9: austenitic stainless steel, 10: twinning-induced plasticity

Much might be learned from other fields: pressure strengthening (cold-stretching) is commonly used to increase the allowable stresses and thus reduce the weight of cryogenic pressure vessels which are from austenitic stainless steels even though the main cause of the yield-strength improvement is unclear. The effect of 9% pre-strain and low temperatures (-196C) upon the cryogenic mechanical properties of S30408 stainless steel was therefore investigated, including the effect of pre-straining upon welded joints. The engineering significance of the pressure strengthening of a cryogenic inner vessel was explored[163] using finite-element analysis. This showed that, at -196C, the strengthening effect of low temperatures played the main role in producing strength increments. The 9% pre-strain could improve the uniformity of the distribution of Vickers microhardness data for welded joints. The prior use of pressure strengthening of inner vessels helped to prevent further plastic deformation during subsequent hydraulic tests, to avoid thermal insulation degradation of vacuum jackets and to even-out stress concentrations in the roots of pipes.

There is an increasing interest in the use of very-high and ultra-high strength materials in lightweighting. The dual-phase and transformation-induced plasticity types are popular advanced high-strength steels, especially with regard to vehicle parts that are involved in crash-energy control. The dual-phase types are favoured for use in manufacturing components for front-end and exterior-panel applications.

The use of dual-phase and transformation-induced plasticity steels results in weight-savings of the order of 10 to 25%. The possible weight-savings become much greater when ultra-high strength steels are used; yielding weight-reductions of the order 30 to 40% when the steels have strengths of 1300 to 1500MPa. Strength levels of the order of 1800 to 2000MPa are envisioned for the production of intrusion-resistant parts, with an ultra-high strength, good stiffness and very little (< 5%) collision-related deformation being specified for those parts.

Hot press-formed 22MnB5 steels are commonly used for B-pillar and front–rear reinforcement applications, with adequate strength sometimes being achieved by sacrificing formability. The mechanical properties of transformation-induced plasticity steels can be adjusted by modifying the strain-hardening step. It is interesting to compare the mechanical properties of 1GPa transformation-induced plasticity steel with those of the normally employed ferritic and austenitic steel sheet. A transformation-induced plasticity steel possesses twice the uniform elongation, and a much higher ultimate tensile strength (figure 15).

An unusually-high strain-hardening behaviour leads to a better press-forming performance, and is attributed to a dynamic Hall-Petch effect. The predominant deformation mode in transformation-induced plasticity steel is dislocation glide, but the dynamic Hall–Petch effect is associated with the continuous formation of mechanical twins during straining. The formation of the mechanical twins involves the creation of new crystal orientations and the twins then gradually reduce the available glide distance of dislocations and increase the flow stress. The gradual increase in twin density then results in the very high strain-hardening which is typical of transformation-induced plasticity steel.

The twin formation is controlled by the stacking-fault energy. If the stacking-fault energy is 20mJ/m^2, strain-induced transformation is more likely to occur. Twin-formation is suppressed if the stacking-fault energy is $50 mJ/m^2$. Austenitic transformation-induced plasticity steels with manganese contents of between 12 and 30wt% are of particular interest. They have strength x ductility products ranging from 40000 to 60000MPa%. The other alloying additions can include carbon, aluminium and silicon. The carbon, manganese, aluminium and silicon contents are chosen so as to lead to stacking-fault energies of 20 to $50 mJ/m^2$. Carbon additions of 0.4 to 1.0wt% stabilise the austenite phase and produce solid-solution strengthening.

It was originally feared that transformation-induced plasticity steel would be prone to delayed fracture and dynamic strain-aging, but it was then discovered that aluminium additions suppressed delayed fracture. Three main types of transformation-induced plasticity steel are: Fe–(25 to 30)Mn-3Si-3Al, Fe-22Mn-0.6C and Fe-18Mn-1.5Al-0.6C. A marked improvement in the mechanical properties of transformation-induced plasticity steel is found upon reducing the manganese content, thus favouring the use of 15wt% of manganese; for example, Fe-15Mn-2.0Al-0.6C-0.5Si, Fe-15Mn-3.0Al-0.7C, Fe-15Mn-2.5Al-0.7C, Fe-15Mn-2.5Si-2.5Al-0.6C, Fe-15Mn-2.5Al-0.6C and Fe-16Mn-0.6C-0.2Si-0.2Al. Other variants include Fe-12Mn-0.8C and Fe-12Mn-2Si-0.9C.

The iron-rich side of the iron-manganese phase diagram contains an open γ-loop but, at between 5 and 25wt% manganese, the microstructure is dominated by the appearance of α'-martensite at low manganese contents and ε-martensite at higher manganese contents. The inclusion of some 27wt% of manganese is required in order to ensure the presence of metastable austenite at room temperature.

The addition of just 0.6wt% of carbon produces a martensite-free austenitic microstructure at manganese contents as low as 12wt%. The microstructure of transformation-induced plasticity steel is typically single-phase austenitic in form and comprises relatively coarse grains which frequently contain wide recrystallisation twins. Aluminium is added in order to control the stacking-fault energy of transformation-induced plasticity steel. It also suppresses the formation of Fe_3C. Aluminium additions of less than 3wt% also result in a slight reduction in density, due to the combined effect of its lower molecular weight and a related increase in the lattice parameter.

A transformation-induced plasticity steel having a uniform carbide-free austenitic microstructure can also be created by using a higher manganese content and avoiding the addition of carbon, while using silicon and aluminium additions to control the stacking-fault energy. The aluminium is essential because it greatly improves the transformation-induced plasticity properties: small additions of aluminium augment the transformation-induced plasticity effect, and the formation of ε-martensite is effectively suppressed by the addition of 1.5wt%Al to Fe-15Mn-0.6C transformation-induced plasticity steel. The addition of nitrogen has similar effects to those of aluminium.

As noted elsewhere, the stacking-fault energy is a key factor in the control of the mechanical properties of high-manganese alloys, with the best results corresponding to stacking-fault energies of 20 to $50mJ/m^2$. This specific range has been linked to suppression of the athermal $\gamma \rightarrow \epsilon$ martensitic transformation. The stacking-fault energy is, in turn, proportional to the difference in the free energies of the face-centered cubic and hexagonal close-packed structures.

Contributions which arise from the interfacial surface energy and the magnetic energy, associated with the $\gamma \rightarrow \epsilon$ transformation can also have a marked effect and thus have to be taken into account. The interfacial energy can be taken to comprise the coherent twin-boundary energy and the energy of the twinning dislocations. Stable fully-austenitic microstructures having associated transformation-induced plasticity properties, also tend to have stacking-fault energies ranging from 20 to $30mJ/m^2$. Carbon additions are required in order to obtain a low stacking-fault energy. Study[164] of the effect of carbon in Fe-22Mn-C alloys shows that the stacking-fault energy of Fe-22Mn alloy is about $30mJ/m^2$, and that carbon additions of less than 1wt% reduce the stacking-fault energy to about $22mJ/m^2$. The stacking-fault energy increases at higher carbon contents[165]. A stacking-fault energy that is greater than about $25mJ/m^2$ is expected to result in the twinning of a stable γ-phase, while a stacking-fault energy which is smaller than about $16mJ/m^2$ results in ϵ-phase formation. Another study[166] suggested that the stacking-fault energy should be at least $19mJ/m^2$ in order to permit mechanical twinning; a stacking-fault energy of less than $10mJ/m^2$ resulted in ϵ-phase formation. Twinning tends to disappear below a stacking-fault energy of $18mJ/m^2$, and is replaced by ϵ-platelet formation[167]. A stacking-fault energy of $20mJ/m^2$ leads to the best hardening rate.

Aluminium increases the stacking-fault energy and suppresses the $\gamma \rightarrow \epsilon$ transformation. A stacking-fault energy of $33mJ/m^2$ is required in order to provoke twinning in Fe-18Mn-0.6C-1.5Al. The stacking-fault energy of Fe-18Mn-0.6C-1.5Al transformation-induced plasticity steel is $30mJ/m^2$. The stacking-fault energy of Fe-(20.24 to 22.57)Mn-(2 to 3)Si-(0.69 to 2.46)Al alloy, containing 100ppm of carbon and 0.011 to 0.052% of nitrogen, increases with increasing nitrogen content while the likelihood of stacking-fault formation decreases. Aluminium increases the stacking-fault energy by between 1.4 and $5mJ/m^2$ per 1wt% of aluminium. Silicon also increases the stacking-fault energy by about $1mJ/m^2$ per 1wt% of silicon.

A stacking-fault energy change of just 5 to $10mJ/m^2$ can cause a sharp transition from strain-induced martensite formation to strain-induced twinning, and there is often no clear relationship between the stacking-fault energy and the twinning mechanism. It is possible that martensite formation and mechanical twinning are controlled by extrinsic stacking-faults and intrinsic stacking faults, respectively. Deformation of two-phase Fe-19.7Mn-3.1Al-2.9Si steel at 86 and 160C led to ε-martensite and twinning at the lower temperature and to mechanical twinning alone at the higher temperature. Only ε-martensite was observed at room temperature. This has been attributed to the presence of extrinsic stacking faults, at lower temperatures, which were precursors to ε-martensite formation while intrinsic stacking faults were the twin precursors at higher temperatures.

More directly pertinent to the use of these materials in lightweighting is their mechanical properties. The Fe-Mn alloys are characterised by the occurrence of strain-induced martensite formation, deformation twinning, pseudo-twinning, extended dislocation glide and perfect-dislocation glide. The predominant deformation mode is dislocation-glide. The dislocation-density increases at low strains, with the grain boundaries being particularly effective sources of isolated stacking faults. The onset of twinning meanwhile requires the occurrence of multiple slip within deformed grains. The strain-induced twins exhibit a high aspect ratio and span the entire grain. At 20% strain, the high dislocation density between the deformation twins shows that the twin boundaries act as effective barriers to dislocation movement. The twins are very thin, and continuous nucleation of new deformation twins of decreasingly smaller size can occur. As a result, the twin volume-fraction does not make up a large portion of the total volume. The rate of dislocation accumulation automatically increases when an alloy has a low stacking-fault energy, independently of twin formation, because a larger dissociation width more effectively reduces cross-slip and results in a higher rate of dislocation accumulation.

Although these steels exhibit high strain-hardening, they also have a relatively low yield stress. The production of a smaller grain size, and solid-solution hardening, can lead to a higher yield strength. The grain-size dependence of the yield strength obeys the usual Hall- Petch relationship. In the case of Fe-Mn-Al-C transformation-induced plasticity steel, the yield strength is increased by aluminium solid-solution hardening. A decrease in the tensile strength of aluminium-containing transformation-induced plasticity steel can be due to a reduction in strain-hardening that results from a decrease in the rate of strain-induced twin-formation.

The marked work-hardening of these steel is not generally attributed to the effect of the stacking-fault energy. Explanations which are based upon the evolution of the dislocation density and twin volume predict isotropic hardening. This is consistent with the facts that the total volume fraction of twins is very low, and that plastic deformation occurs mainly via dislocation glide. In the first stage, nm-thick twins are assumed to move until they reach a strong boundary. The twins then thicken during a second stage. Two twinning systems operate, in which twins of the first system develop across an entire grain while much shorter and thinner twins of the second system develop between the primary twins.

During the strain-hardening of Fe-18Mn-0.6C-1.5Al transformation-induced plasticity steel, the amount of twinned volume was found to be controlled by an increase in the number of new deformation twins, rather than by their lateral growth. An elasto-viscoplastic model for Fe-22Mn-0.6C transformation-induced plasticity steel indicated that the twinned volume fraction depended upon the grain orientation and amounted to less than 0.08 at a macroscopic strain of 0.4.

The strain-hardening can be calculated on the basis of the evolution of the coupled population-densities of mobile dislocations and immobile forest dislocations. The work-hardening of 980MPa Fe-Mn-Al-C transformation-induced plasticity steel can be almost isotropic up to a longitudinal plastic strain of 0.24. The strain-hardening steel is however very unlikely to be isotropic over the entire deformation path, given that there are indications that kinematic hardening occurs at larger strains.

A Bauschinger effect in these steels has been attributed to the simultaneous deformation of grains and their twins. This implies that a forward internal stress acts on the twin while a backward internal stress acts on the untwinned matrix, because both the matrix and the twins must have similar strain components. The back-stress has been linked to the occurrence of dislocations of a given slip system being stopped at the twin boundaries and building up a stress which prevents similar dislocations from moving. Reducing the manganese content from 18 to 12% results in fracture occurring at lower stresses. An increase in the carbon content to 0.7wt%, plus the addition of 1wt%Al, leads to an appreciable recovery of the original ultimate strength. Further increases in elongation result from the addition of copper.

A transition from slip-only deformation, to slip plus twinning deformation, occurs when the slip stress attains the twinning stress. The latter increases with increasing stacking-fault energy, and the stress which is required to nucleate a twin is related to the intrinsic stacking fault energy via a quadratic or linear relationship. The grain size plays a role in determining the magnitude of the twinning stress, and larger grains tend to widen the twinning domain. In transformation-induced plasticity steel, twin nucleation is not homogeneous in the nucleation stage and the twin formation is closely related to any previous dislocation activity. Strain-induced twin formation occurs only after some degree of prior dislocation generation and dislocation–dislocation interaction has occurred. Twins nucleate in special dislocation configurations that are created by the above interactions, and generally generate multilayer stacking faults which act as twin nuclei.

Table 25. Room-temperature stacking-fault energies of TWIP steels

Composition (wt%)	Phases	Mechanisms	SFE (mJ/m^2)
19Mn-5Cr-1Al-0.25C	γ+ε	twinning	20.90
19Mn-5Cr-2.5Al-0.25C	γ	twinning, glide	30.50
19Mn-5Cr-3.5Al-0.25C	-	-	39.40
19Mn-5Cr-4Al-0.25C	γ+α	twinning, glide	47.50
25Mn-0.6Al-0.15C	γ+ε	twinning, glide	7.75
25Mn-1.5Al-0.15C	γ	-	10.67
25Mn-2.2Al-0.15C	-	-	15.12
25Mn-3.1Al-0.15C	-	twinning	14.95
25Mn-4.8Al-0.15C	-	-	54.74
18Mn-1.5Al-0.6C	-	-	26.40
18Mn-0.6C	-	-	13.00
31Mn-0.17C	-	glide	17.53

Three types of strain-induced twinning-nucleation mechanisms have been proposed. These are pole mechanisms, glide dislocation sources and cross-slip mechanisms. In the pole mechanism, a jog is created on a dislocation due to dislocation intersection. The jog then dissociates into a sessile Frank partial dislocation and a Shockley partial dislocation. When the partial dislocation is moved by an externally applied force, it trails an intrinsic stacking fault and repeatedly rotates around the pole dislocations, thus generating a twin.

Glide-type deformation induced twinning mechanisms have also been proposed. Glide sources are less probable sources of twins, but a transmission electron microscopic study of twinning in Fe–22Mn–0.5C transformation-induced plasticity steel supported a model for the creation of a three-layer stacking fault acting as a twin embryos. These are formed by the interaction of two coplanar glide dislocations and the interaction led to the formation of an extrinsic stacking fault configuration. Larger twins were formed by the growth of twin embryos into each other. In cross-slip models for twinning, a twinning partial is formed via a double cross-slip mechanism or by the interaction of a primary dislocation with a Frank dipole, a faulted dipole or a Lomer–Cottrell lock.

Although mechanical twins could in principle be nucleated at dislocations, grain boundaries or other pre-existing defects, it appears that some pre-deformation, considerable dislocation interactions and the nucleation of extrinsic stacking faults are always required in order to nucleate twins. The grain orientation and crystallographic texture can, in general, also have a marked effect upon twinning. For example, whereas 100nm oriented grains which are deformed in tension do not twin, grains which are oriented with an axis parallel to the tensile axis are heavily twinned.

Table 26. Stacking-fault energy of Fe-22Mn-0.6wt%C TWIP steel as a function of temperature

Temperature (K)	Phase	Mechanism	SFE (mJ/m^2)
77	γ	glide	10.00
293	γ	glide, twinning	19.00
673	γ	glide	80.00

When compared with an uniaxial tensile test, deformation in press forming can be very complicated, involving complex and inhomogeneous conditions of strain and stress. The yield surface is often needed for numerical simulations and the shape of the yield surface is known to have a significant effect on the forming limit of anisotropic sheet metals. In addition, typical values for the angular dependence of the normal anisotropy are listed. The planar anisotropy of transformation-induced plasticity steel is limited. In sheet forming, the forming limit diagram is obtained by determining experimentally the strains at which necking occurs for different deformation paths. The necking strain in the plane strain deformation mode, the forming limit curve minimum value of transformation-induced plasticity steel is very high (45%) compared to those for 590MPa dual phase steel (30%) and 780MPa TRIP steel (28%). The press-forming properties of transformation-induced plasticity steel (tables 25 and 26) have therefore proved to be excellent as illustrated by the example of a shock-absorber housing. This makes transformation-induced plasticity steel ideally suited for the press-forming of high-strength parts having a complex shape. The stretch flanging properties of transformation-induced plasticity steel, typically evaluated by means of a hole expansion test, are considerably better than those of other advanced high-strength steels of similar strength level, but are not as good as those of ferritic deep-drawing steel grades. The combination of low normal anisotropy and low strain-rate sensitivity results in lower hole expansion ratios. This is very pronounced when less than optimum hole edge preparation leads to considerable hole-edge deformation, as in the case of hole-punching. The dynamic energy-absorption of different types of automotive steels when tested as a function of strain-rate have been compared. The high strain-rate testing of Fe–25Mn–3Si–3Al–0.03C transformation-induced plasticity steel showed that extensive twinning occurred during high-strain rate deformation, but no brittle fracture was observed, even at low temperatures.

A study of the high strain-rate deformation of Fe–31Mn–3Si–3Al transformation-induced plasticity steel with a grain size ranging from 1.1 to 35.5mm showed that, unlike the case for ferritic steels, there were still large elongations for small grain sizes. A study of the high strain-rate behaviour of Fe–24Mn–0.5Si–(0.11–0.14)C transformation-induced plasticity steels which contained 0.91 or 3.5% of aluminium found that the transformation of austenite to martensite occurred up to a given strain rate. Transformation-induced plasticity steel which contained 3.5% of aluminium exhibited greater stability and transformation of this transformation-induced plasticity steel was limited in strain-rate range.

Regardless of the aluminium content, the transformation of the austenite phase was suppressed during high strain-rate deformation due to adiabatic heating of the sample. On the basis of observations of serrated grain boundaries, it has been suggested that dynamic recrystallisation could be taking place during high strain-rate tests. Dynamic strain-aging occurs at room temperature in Fe–22Mn–0.6C and Fe–18Mn–0.6C transformation-induced plasticity steels, and this is revealed by serrations in the stress–strain curves. During dynamic strain-aging the mechanical deformation of transformation-induced plasticity steel is localised entirely into deformation bands which traverse tensile-test specimens. The properties of these bands reflect fluctuations of the strain-state at two points in the tensile sample during motion of the Portevin–LeChatelier bands. The properties of the latter bands have been analysed in detail, showing that the band velocity decreases with strain and that the band strain-rate is 15 to 100 faster than the applied value. Localisation can lead to press-forming difficulties, but the occurrence of Portevin–LeChatelier bands during uniaxial tensile testing is not known to lead to poor press-forming results for transformation-induced plasticity steel. This is suggested to be due to the relatively high strain rates which are used during press-forming.

Other aspects of dynamic strain-aging should not be ignored, because it results in a negative strain-rate sensitivity and causes limited post-uniform elongation. In the case of carbon-alloyed face-centered cubic alloys, the room-temperature dynamic strain-aging cannot be explained in terms of the long-range diffusion of carbon. It instead results from the short-range order which is due to the presence of point-defect complexes that can re-orient themselves within the stress-field of dislocations or in the stacking-faults.

Among the potential defect complexes in transformation-induced plasticity steels are a C–vacancy complex, a C–C complex and a C–Mn complex. The most likely complex contains a single carbon atom in an octahedral interstice and one manganese atom. Direct evidence for the presence of such point-defect complexes, and for their interaction with dislocations, is deduced from internal friction data. Re-orientation of the point-defect complex does not require long-range diffusion, and only a single diffusional jump of the interstitial carbon atom in the complex is required in order to take up a suitable orientation with respect to the strain field of the partial dislocation or the stacking fault

Dynamic strain-aging and its associated serrated stress–strain curves can be avoided by adding aluminium. As the content of the latter atoms increases the stacking-fault energy, but does not interact with the point-defect complex, and decreases the carbon diffusivity in transformation-induced plasticity steels, the results suggest that the interaction which gives to flow-localisation is the interaction between the C–Mn point-defect complexes and the stacking-faults.

Manganese-rich octahedra are preferred locations for carbon atoms and the probability of a pure manganese octahedron is low. Meanwhile the Mn–C bond is stronger than the Fe–C bond. It is suggested that the Fe–Fe and Mn–Mn bonds are affected differently by the presence of carbon. In the presence of carbon the Fe–Fe bond strength is slightly destabilised and the Fe–Fe bond-length slightly increases while the Mn–Mn bond-length decreases. This leads to stabilisation of carbon- containing manganese-rich octahedra and to short-range ordering.

Delayed fracture was once thought to be a major problem for Fe–22Mn–0.6C transformation-induced plasticity steel, with the effect appearing in fully deep-drawn cups shortly after drawing. High residual tensile hoop-stresses are known to exist at the edge of fully-drawn cups. It has been proposed that it is related to martensitic transformation in the presence of residual stresses and hydrogen. This was investigated with regard to the influence of martensite which formed during the deformation of Fe–18Mn–0.6C and Fe–18Mn–0.6C–1.5Al transformation-induced plasticity steel. Aluminium-containing transformation-induced plasticity steel was free of martensite following tensile testing but both types of transformation-induced plasticity steel contained martensite following cup-drawing. The amount of embrittling martensite was lower in aluminium-containing transformation-induced plasticity steel.

A comparison was made of the hydrogen-embrittlement of twinning-induced plasticity and transformation-induced plasticity steels following cathodic hydrogen-charging. It was reported that Fe–15Mn–0.45C–1Al and Fe–18Mn–0.6C twinning-induced plasticity steels, with and without aluminium, contained less hydrogen and were much more resistant to embrittlement than were transformation-induced plasticity steel following U-bend and cup-drawing tests. The delayed fracture problem prompts much interest in testing the sensitivity of twinning-induced plasticity steel to hydrogen-induced embrittlement. The absence of a noticeable degradation of the mechanical properties following hydrogen-charging, apart from a relatively small reduction in the total elongation, suggests that transformation-induced plasticity steel may be quite insensitive to hydrogen-induced fracture. Delayed fracture in transformation-induced plasticity steel is suggested to be a form of stress corrosion cracking given that it seems to require the confluence of a high residual stress, a strong hydrogen trap-site and hydrogen-absorption. Delayed fracture does not seem to occur if any of those factors is absent.

In the as-received state transformation-induced plasticity steels exhibit cyclic softening during fatigue testing. The dislocation density decreases during cyclic tests and existing twins widen. A resultant lack of dislocation-twin interactions and of new-twin nucleation are the main causes of softening. Pre-deformation improves the fatigue life of Fe–22Mn–0.52C twinning-induced plasticity steel. The stress–strain curves of transformation-induced plasticity steel reveal the suppression of serrations due to dynamic strain aging by adding aluminium. Twins which form during pre-deformation lead to stable cyclic loading and a longer fatigue life. High-cycle bending fatigue tests of Fe–22.3Mn–0.6C, Fe–17.8Mn–0.6C and Fe–16.4Mn–0.29C–1.54Al steels indicated a 400MPa fatigue stress limit; well above the yield stress.

Nanostructured Fe–22Mn–0.6C twinning-induced plasticity steel, obtained by cold-deformation and recovery annealing has a very high yield strength and an adequate elongation. The preparation process decreases the dislocation density and retains a very dense nanoscale twin microstructure. A considerable interest in high-manganese twinning-induced plasticity steel steels is due to their superior mechanical properties. As compared with conventional low-carbon steels, high-manganese twinning-induced plasticity steels have high carbon and manganese contents. When aluminium is added, its content also tends to be high.

The production of high-manganese twinning-induced plasticity steel originally encountered many technical difficulties. One of these was the ferromanganese alloy required for the steel-making process. Standard ferromanganese, with its high phosphorus content, is unsuitable for the production of good-quality transformation-induced plasticity sheet steel products. Alternative working methods are also required for the production of transformation-induced plasticity steel. These include pre-melting of the ferromanganese additions before alloying, and the use of liquid rather than powder fluxes during continuous casting in order to ensure edge quality and avoid cracking. Oxide scale formation during reheating should be controlled in order to avoid internal grain-boundary oxidation and resultant surface defects or edge cracking.

Cold-rolling reduction is slightly limited due to the high strain-hardening behaviour of transformation-induced plasticity steels. During recrystallisation annealing following cold-rolling, recrystallisation requires 1000s at 560C and only 10s at 630C.

One approach to lightweighting design is based upon the use of fatigue accumulated damage theory to improve reliability by means of optimum structural design. Another approach is to use high-performance lightweight materials. The former method takes little account of the structural effects caused by massive small loads during service, but the potential strength of the component is not fully exploited. In the case of the latter method, cost is the major criterion. In order to apply lightweighting to an electric (fuel-cell) vehicle, the drive shafts were re-designed[168]. The method was based upon low-amplitude load-strengthening of the material and permitted the stress which corresponded to the test load to be within the strengthened range of the material. Under these conditions, the lightweighting design was expected to be able to ensure that the reliability of the shaft was not reduced, and even maximize the potential strength of the part in order to achieve weight reduction and consequently cost. The feasibility of the design was verified by strength and modal analyses, based upon computer-aided design, of a lightweighted shaft. The design was applied to the loading of the half-shaft of an independent axle.

It is to be noted that heavy-duty vehicles can benefit from the lightweighting strategies developed for cars, but the parts must be then be designed so as to take a greater account of fatigue-resistance, especially around trimmed areas, and of stiffness. It is found that surface treatments such as sand-blasting or shot-peening help to increase fatigue life of advanced high-strength steels in trimmed areas and thus permit weight-reduction via sheet-gauge down-sizing. Any resultant decrease in stiffness due to the thickness reduction can be compensated by design changes.

In order to explore a reduction in the weight of the transmission components of vehicles, time-domain loading data were harvested[169] from a transmission system under typical working conditions. These provided information for load-spectrum synthesis and generated a programmed load-spectrum which consisted of various amplitudes. The load-spectrum could be used for fatigue loading of the output flange of a gear-box. A finite-element model of the flange was then set up and stress analysis was performed at stress-concentrators such as a fillet. Following fatigue-life predictions based upon the compiled load-spectrum and the theory of cumulative fatigue-damage, the fatigue life of the outside fillet of the flange - a site of maximum stress - was obtained. It was found to be possible to obtain useable fatigue-life predictions by using load-spectrum, finite-element analysis and stress-life approaches to fatigue-damage calculation.

Press-hardened steel coated with aluminium-silicon alloy have been used in vehicle-body structures in order to combine lightweighting with crashworthiness. The weldability of this material is unfortunately quite poor due to its high carbon-equivalence and high strength, plus severe associated heat-affected zone softening and interfacial fracture. The interfacial fracture mode is partly due to the presence of retained aluminium-silicon coating as a sharp notch at the faying interface following welding. This impairs the mechanical performance of the joint. In order to improve the weldability of press-hardened steel, a stepped current-pulse method was proposed[170] for the resistance spot welding of 1.5mm-thick aluminium-silicon coated press-hardened steel. This was intended to promote nugget-growth, to inhibit expulsion and to eliminate any residual aluminium-silicon layer. It succeeded in markedly improving the mechanical performance of joints and promoting the desired button pull-out fracture mode.

The effect of low-temperature tempering upon the microstructure and properties of medium molybdenum-content martensitic steel was studied[171] in order to determine the influence of paint-baking during vehicle manufacture. There was a marked improvement in the tensile ductility, which was associated with a change from brittle to ductile fracture. The ductilization was in turn attributed to carbon redistribution during baking and to the high (15%) volume fraction of retained austenite. The baked medium-manganese martensitic steel offered an excellent combination of strength and ductility, thus making it a good choice for vehicle-lightweighting use. Similar results have been found for bainitic ferrite and dual-phase steels which were heated at 120 to 200C, for 0.25 to 1h, after pre-straining to 0, 2 or 5%. The bake-hardened results for dual-phase and bainitic ferrite ranged from 90 to 140MPa, after pre-straining to 2% and baking (170C, 1200s). The ductility was lower following 5% pre-strain, and the elongations could fall to 1%. Press-hardened steels instead exhibited marked increases in yield strength, even in the absence of pre-strain, that amounted to 122 or 175MPa following baking (160C, 1h). On the other hand, the ultimate tensile strengths decreased due to a reduced strain hardening and the total elongations also decreased slightly.

Titanium

Titanium use has long been tipped to become the third wave of light metals to be used, following aluminium and magnesium[172] ... but has seemingly yet to occur. Titanium and its alloys offer a unique combination of properties, but suffer from a high raw-material cost. Life-cycle assessment can be used to identify novel routes, to metal production and product manufacture, that are more energy-efficient and less expensive to set up, while using ecologically favorable material supply-chains. One route to titanium production is the initial mining of ilmenite, followed by upgrading of the ilmenite to synthetic rutile and titanium-metal separation using the usual Kroll process. Alternative processes for titanium-metal production involved direct electrochemical de-oxygenation of titanium dioxide, or the plasma-processing of titanium tetrachloride. The environmental effects which were included in the life-cycle assessment were greenhouse-gas emission and acid-rain production. It was concluded that, in spite of the relatively large amounts of primary energy which were required for the production of the light metals, their corrosion-resistance, strength and density helped to outweigh the environmental effects.

An unsprung suspension component was selected and re-designed from the viewpoint of replacement with titanium. During the re-design, the manufacturing procedure and processing were integrated. Following design and analysis, components were prototyped and subjected to both static laboratory testing and in-vehicle dynamic testing. This was a promising example of a titanium component being produced for a current production vehicle. The costs were also such as to justify its fabrication[173]. Other considerations included adaptation to high-volume production and the availability of near-net size plate and bar stock. The fuel-cost savings and consequent life-cycle costs were compared with life-cycle duration and unit fuel-price in order to identify profitable operating conditions. It was found that the total life-cycle costs were quite comparable to those of the original component, thus justifying the replacement[174].

Joining

The trend to lightweighting naturally leads to the design of multi-material structures which typically involves joining aluminium to steel. A new resistance spot-welding technique[175] uses a multi-ring domed electrode and various solidification rates and, in the case of aluminium and steel, an iron-aluminium intermetallic compound layer is formed at the interface. The strength of this compound then influences the tensile-shear fracture mode: interfacial or pull-out. On the basis of experimental heat-affected zone and intermetallic compound shear strengths, it was concluded that it was not adequate to use the recommended weld-nugget diameter in order to determine the fracture modes of aluminium-steel spot welds. A new formula which took account of the shear strength of the intermetallic layer in aluminium-steel resistance spot-welds was developed in order to calculate a critical weld-nugget diameter on the basis of experimental results. Most current weld-quality acceptance criteria are empirical and have traditionally been deemed to be adequate. The rush to accelerate lightweighting has however led to new forms of joint; for instance, those between dissimilar materials. These may not conform to traditional acceptance criteria. The latter can be too conservative in some cases and not conservative in others when considering thin-gauge and lightweight components. This has been likened[176] to the situation experienced with additively manufactured metallic components. Those can be viewed as being welded components which contain randomly distributed discontinuities throughout their volume.

Resistance spot welding is the main joining method for automotive sheet steel. Such welding of transformation-induced plasticity steel is more difficult than for plain carbon steels. Because transformation-induced plasticity steel has a fully-austenitic microstructure, there is a danger of solidification cracks. Twinning-induced plasticity steel has a smaller welding-current range and weld expulsion occurs at low welding currents. The weld bead is lower in hardness than the base metal. Common transformation-induced plasticity steel weld defects include grain-boundary liquation cracking in the heat affected zone and voids.

The fatigue, noise and vibration of a vehicle depend greatly upon its stiffness and the framework of the automobile body comprises various beam-like structural components that are connected by spot-welds or rivets at the flanges or brackets. In addition to their cross-sectional forms, their supports and connections markedly affect the effective beam stiffness. The relationship between boundary-support stiffness and actual stiffness was analyzed[177], and the bending and torsional characteristics of thin-walled beams having various cross-sections were compared. Although they offer improved crashworthiness and lightweighting, advanced high-strength steels can compromise the stiffness, noise and vibration aspects of vehicles; due largely to the reduced part-thicknesses. In order to improve the torsional rigidity of a chassis frame, bulkhead pairs can be added as reinforcement; thus increasing the thickness of frame parts. A material efficiency ratio can be used to evaluate the efficiency of a weight-increase in improving the stiffness, noise and vibration characteristics of a vehicle. The addition of a bulkhead pair gives the highest material efficiency ratio, but has a limited potential for improving torsional rigidity. On the other hand, increasing part thickness and enlarging the closed sections on rails imparts a higher torsional improvement, but the material efficiency ratio is much lower. There is a degree of synergy involved here in that the addition of bulkhead pairs, together with enlargement of the rail sections, produces a greater material efficiency and a greater potential rigidity improvement than does either of the individual methods.

The mechanical behavior, under bending loads, of a structure which consisted of a composite hat-section that was attached to a metal sheet has been examined[178]. Three attachment methods were evaluated: riveting, adhesive bonding and their combination.

Quasi-static tests were performed at room temperature and at -40C. No joint failure was observed in quasi-static tests, but the attachment technique affected the stiffness and failure-mode of the structure. Some specimens suffered appreciable joint-damage during fatigue, followed by catastrophic failure. Finite-element models could predict the behavior of the structure up to damage of the composite[179]. It was concluded that the use of dissimilar materials and effective joining techniques would lead to a reduced vehicle weight and lower fuel consumption.

Table 27. Weld strengths of magnesium alloys

Alloy	Temper	YS (MPa)	UTS (MPa)	Weld Type	Strength*(%)
AZ31B	H112	160-200	280-320	MIG	94
AZ31B	H112	160-200	280-320	TIG	95-97
AZ31B	H112	160-200	280-320	FSW	92
AZ80	T5	267	350	TIG	72
ZK60	T6	305-350	365-410	FSW	94
Mg-Zn-Y-Zr	T6	321	356	FSW	93
AZ91	O	280	331	TIG	80

*ratio of weld-strength to base-metal strength

Joints which are made by using direct resistance spot welding to join aluminium alloy to steel tend to suffer from poor mechanical properties, due to the formation of thick brittle intermetallic compounds and solidification-related defects. An ultrasonic-plus-resistance spot-welding technique was used[180] to join 1mm-thick AA6061-T6 to 0.9mm-thick AISI1008 steel, leading to a joint strength of 3.7kN and a button pull-out failure mode. The morphology and thickness of the intermetallic layer were characterized by means of scanning electron microscopy, showing that a layer of less than 1.4pm of intermetallic compounds formed at the aluminium|steel interface. The ultrasonic plus resistance spot welding method has also been used[181] to join AA6022 aluminium alloy to zinc-coated DP980 dual-phase steel. During the solid-state ultrasonic spot welding, an interface structure which comprised multilayer aluminium-zinc and zinc-iron intermetallic compounds formed due to alloying of aluminium with the steel coating. Such structures then melted into the aluminium nugget and new aluminium-iron intermetallics formed during resistance spot welding. The ultrasonic plus resistance spot-welded joints exhibited a higher fracture energy than that of direct resistance spot-welded joints. An electrospark-deposited AA4043 interlayer has been used for the resistance spot-welding of AA5052 aluminium alloy to galvanized GI DP600 dual-phase steel. A tensile lap-shear strength increase of at least 30% was obtained by using the interlayer. This technique is expected to permit the greater use of aluminium alloys in vehicle lightweighting.

The surface oxide which is ubiquitous on aluminium-alloy components poses however a considerable barrier to the use of traditional resistance spot-welding. Self-piercing riveting has thus become a popular alternative joining method. In order to permit the use of resistance spot-welding to join steel to aluminium, an electrode has been developed[182] which has weld-face designs that include sharp radii-of-curvature and a surface topography which reduces the effects of the oxide layers on aluminium alloys by reducing the contact-resistance. Microstructural analysis and mechanical testing revealed the effect of electrode weld-face topographies upon the resistance spot-welding of 1.2mm-thick 6022-T4 aluminium sheet; welded to itself. Three types of electrode-face design were compared: a textured electrode and a multi-ring domed electrode which could have two possible ring-heights. It was found that the electrode-face topography could markedly affect resistance-heat generation and electrode-cooling.

Multi-ring domed electrodes have been used[183] to spot-weld 0.8mm-thick X626-T4 aluminium-alloy sheet. The resultant spot welds had an equiaxed grain structure around the weld-nugget periphery while, within the weld-nugget, fine columnar grains were visible at the nugget perimeter. Welding defects were concentrated at the centre of the nugget. By using modified shear specimens, it was found that the weld-nugget was the weakest location whereas the heat-affected zone is improved by a precipitation-aging that was due to a combination of welding and paint-baking thermal cycles. A new equation was again developed for calculating the minimum weld-nugget diameter to ensure that interfacial fracture does not occur during tensile and fatigue testing. Load-controlled fatigue tests showed that fatigue-life curves for both lap-shear and coach-peel configurations fell on a single master-curve. This indicated that the weld-nugget size was the factor which controlled fatigue life. The fracture-mode depended upon the load-amplitude, and ranged from full button pull-out at high amplitudes to eyebrow-like fracture with kinked crack paths at low amplitudes. No interfacial fracture occurred during fatigue-testing, due to the large size of the weld-nugget; even though the latter was the weakest location.

Table 28. Weld strengths of aluminium alloys

Alloy	Temper	YS (MPa)	UTS (MPa)	Weld Type	Strength*(%)
AA5083	H111	125-200	275-350	TIG	90
AA5083	H111	125-200	275-350	FSW	93
AA5083	O	148	298	FSW	100
AA5083	H321	153	305	FSW	91
AA6005A	T6	200-225	250-270	MIG	75
AA6061	T6	270	290	MIG	75
AA6061	T6	270	290	TIG	64
AA6061	T6	270	290	FSW	84
AA6082	T6	250-260	290-310	MIG	78
AA6082	T6	250-260	290-310	TIG	67
AA6082	T6	250-260	290-310	FSW	83
AA6082	T4	149	260	FSW	93
AA7018	T79	245	350	FSW	95
AA7020	T6	348	395	MIG	77
AA7020	T6	348	395	FSW	78

*ratio of weld-strength to base-metal strength

Adhesives have been used[184] to bond the 5182 aluminium alloy which is commonly used in vehicle bodies. Various surface treatments were used to change the surface quality of panels, and the effects of factors such as the surface roughness and the microscopic surface area upon the morphology of the joint interface and mechanical properties were studied. The wettability and mechanical properties of the fracture surface were analyzed by making surface free-energy measurements, tensile or shear tests and optical microscopic observations. The use of emery-cloth abrasion could increase the surface roughness and effective bonding-area of aluminium-alloy panels, increase the mechanical gripping effects between adhesive and panel and improve the joint bonding-strength. There was an optimum degree of abrasion which imparted the greatest increase in joint bonding-strength; a higher or lower roughness led to a reduction in the effective contact area between the adhesive and the panel, thus lowering the bonding strength of joints.

Quantitative assessment of the effect of the adherent response upon the ultimate strength of an adhesively bonded joint is desirable for optimum joint design (tables 27 and 28). Thin-adherent single lap shear-testing was carried out[185] by using three of the sheet metals which are typically used to replace mild steel for lightweighting purposes. These were hot-stamped Usibor 1500AS ultra high-strength steel, AA5182 aluminium alloy and ZEK100 magnesium alloy. Six combinations of single and multi-material samples were bonded using a one-part epoxy adhesive. Finite-element models were used to assess quantitatively the effects of the mixed modes upon final joint failure. The steel|steel joint strength of 27.2MPa was much higher than that of all of the other material combinations, for which the joint strengths were between 17.9 and 23.9MPa. The theoretical models predicted the bending deformation of the adherents, which led to mixed-mode loading of the adhesive. The critical cohesive element in the steel|steel simulation predicted the occurrence of 85% mode-II loading at failure, while the predictions for the other combinations gave between 41 and 53% mode-II loading at failure; thus explaining the higher failure strength of the steel|steel joint. This all emphasized the importance of adherent bending stiffness on joint rotation and ultimate joint strength.

Precipitation-hardenable 6000-series aluminium alloys offer a relatively high strength, together with excellent extrudability and corrosion resistance; thus making AA6061 a very good choice for vehicle lightweighting. On the other hand, the weldability of this alloy by using conventional methods is limited by a tendency to solidification-cracking. The use of friction stir welding is a very promising alternative method for assembling aluminium structures, due to the low release of heat which is involved; thus minimizing crack formation and an improved mechanical integrity of assemblies. The friction stir welding of 3.18mm-thick AA6061-T6 sheet in a lap-joint configuration was considered[186] to be more difficult than in the butt-joint configuration. This was due to the orientation of the interface with respect to the welding tools, together with the need to break through the oxide layer on two planar aluminium-alloy surfaces. Unacceptable degrees of material expulsion and/or appreciable thinning of one of the overlapped sheets occurred under most conditions. The characteristics of the metallurgical bonding at the dissimilar-material interface are strongly affected by the welding temperature[187]. Control of the friction stir welding process temperature permits metallurgical bonding with a suppressed formation of intermetallics at the interface that results in improved mechanical properties.

The lap shear-strength and fatigue properties of friction stir spot-welded AZ31B-H24 magnesium alloy and alloy 5754-O aluminium alloy were determined for 3 configurations: Al(top)|Mg(bottom), Al|Mg with an adhesive interlayer and Mg|Al with an adhesive interlayer. For all of the Mg-Al weld combinations[188], friction stir spot-welding introduced an interfacial layer, into the stir-zone, that comprised the intermetallic compounds, Al_3Mg_2 and $Al_{12}Mg_{17}$. This led to an increased hardness. The Mg|Al and Al|Mg adhesive welds both exhibited a much higher lap shear-strength, failure energy and fatigue life than did the Al|Mg weld without adhesive. Two different types of fatigue-failure mode occurred. In the case of the Al|Mg adhesive joint, at high cyclic loads, nugget pull-out failure occurred due to fatigue cracks propagating circumferentially around the nugget. At low cyclic loads, fatigue failure occurred in the bottom magnesium sheet, due to the stress concentration effect of the keyhole. This led to crack-initiation, followed by propagation perpendicular to the loading direction. In the case of the Mg|Al adhesive joint, a nugget pull-out failure-mode was generally observed at both high and low cyclic loads. The Mg|Mg weld had a nugget-shaped stir-zone around the keyhole, where fine recrystallized equiaxed grains were observed[189]. The hardness profile of this weld had a W-shaped appearance. The Mg|Mg and Al|Al welds had a much higher lap shear-strength, failure energy and fatigue life than those of the Al|Mg weld. Although the Al|Al weld had a slightly lower lap shear-strength than that of the Mg|Mg weld, the former had a higher failure energy and fatigue life. Three types of failure mode were observed here. In the case of the Mg|Mg and Al|Al welds at higher cyclic loads, nugget pull-out failure occurred due to circumferential fatigue-crack propagation around the nugget. At lower cyclic loads, fatigue-failure occurred perpendicular to the loading direction and was caused by the opening of the keyhole due to crack initiation in the heat-affected zone. In the case of the Al|Mg weld, a nugget-debonding failure mode was caused by the presence of an interfacial layer of intermetallic compound.

Self-piercing riveting is a cold mechanical joining process which can be used to join two or more sheets of material by driving a rivet through the top sheet and linking the bottom sheet under the guidance of a die. It is currently the best joining method for aluminium and mixed-material lightweight automotive structures. It is not a new method but it has come to prominence because of the increasing need to join lightweight materials such as aluminium to itself, to steel and other mixed-material structures.

The effectiveness of four methods for joining aluminium to aluminium or aluminium to steel were investigated[190]. These were self-piercing (Henrob) riveting, mechanical (Tox) clinching, a combination of Henrob riveting and adhesive bonding (Henrob-bonding) and a combination of Tox clinching and adhesive bonding (Tox bonding). The strengths of aluminium/aluminium joints and aluminium/steel joints were compared with those of 0.8mm steel/steel spot-welds and weld-bonded joints. It was found that the strength of aluminium/steel joints was roughly equal to that of aluminium/aluminium joints. The strengths of 1.6mm aluminium/aluminium joints made by Henrob riveting and of 1.3mm aluminium/aluminium joints made by Henrob-bonding were equivalent to those of 0.8mm steel/steel joints made by spot-welding and those of 1.5mm aluminium/aluminium joints made by Henrob-bonding were equivalent to those of 0.8mm steel joints made by weld bonding. In the case of Tox clinching and Tox bonding, the strengths of less-than 1.6mm aluminium/aluminium joints were lower than those of 0.8mm steel/steel joints made by spot-welding.

When compared with other conventional joining methods, it offers many advantages which have been listed elsewhere. The aluminium alloys which are joined by self-piercing riveting can be wrought, extruded or cast. The grades of wrought aluminium alloy which are used in automotive bodies include 5xxx, 6xxx, etc. The steels which can be joined include mild, high-strength and advanced high-strength types.

The general requirements for a material that can be joined include sufficient ductility, and this is especially so for the material which is next to the die, so that severe cracks will not be generated at the joint button. The material should also have a hardness/strength which is much less than that of the rivet, so that the latter can penetrate the material and have a sufficiently high interlock-distance without creating excessive compression or buckling. Brittle materials can perhaps be joined when used as the upper or middle material, but not when it is the bottom material on the die side.

In a two-layer stack, the ratio between the thickness of the top and bottom materials can influence the rivetability of the stack and the strength of the joint. Better rivetability and strength will exist when a thinner sheet is used as the top material and a thicker sheet is used as the bottom material. Due to access limitations however rivets can sometimes be pierced only from the thicker-sheet side and, in this case, a careful choice of rivets and dies is required in order to obtain the desired joint quality.

A study[191] of the joining of an aluminium alloy to mild steel showed that, in order to use the aluminium alloy as the top sheet and the steel as the bottom sheet, the top sheet had to be thinner than the bottom sheet. Studies[192] of the fatigue performance of aluminium/mild steel joints, with a thickness of 1.5mm for both AA5052 and cold-rolled mild steel, showed that joints with steel as the top sheet had a better static lap shear strength, but joints with steel as the bottom sheet had a better fatigue strength.

Self-piercing riveting is a good process for the mechanical joining of dissimilar materials but, when magnesium-alloy sheet is the bottom layer, cracks always occur due to its low ductility. A study of the feasibility of joining ultra high-strength steel to an aluminium alloy showed that, if the rivet was not hard enough, joint defects arising from the rivets could occur such as rivet fracture, rivet compression and rivet bending. These defects normally occur when the rivet is too soft for the materials which are being joined. By using suitable rivets and dies it was possible[193] to join SPFC980 ultra high-strength steel, with a tensile strength of about 980MPa, to AA5052. Another case of rivet failure due to compression and fracture during riveting process was exemplified[194] by the attempted joining of aluminium alloys by using aluminium rivets.

The joining of magnesium alloys by using self-piercing riveting is always problematic because they have a low ductility at room temperature due to the hexagonal close-packed structure. The ductility increases with temperature however and, following local heating to 280C using induction heating, it was possible[195] to join AZ31 when a top or bottom material. For this reason it has been proposed to employ laser-heating assisted self-piercing riveting to join magnesium alloys. Wrought strips of AZ31B magnesium alloy with thicknesses of 2.35 and 3.2mm were heated to above 200C. The AZ31 could then be successfully riveted without generating cracks[196]. Local heating of the magnesium alloy substrate not only suppresses cracking of the magnesium alloy when it is used as the bottom sheet, but also improves the setting of the rivet-head and promotes interlocking. In order to ensure a sufficient interlock distance between the rivet and the sheet materials and to obtain optimum joint strength however the thicker magnesium alloy substrate needs to be placed on the die-side. Another study[197] of the effect of local heating upon joint strength and failure mode showed that, when riveting 2mm AZ31 to 2mm AZ31 at room temperature, severe cracking occurred at the joint button. When the AZ31 was pre-heated to 180C or above, the cracks were eliminated. By pre-heating the AZ31 the lap shear-strength of the joints could be increased and the failure of joints during lap shear tests changed from tearing of the bottom sheet to the rivets being pulled out of the bottom sheet. Inconsistences in the results on the rivetability of AZ31 magnesium alloy have been attributed to the variable mechanical properties of the alloy. The latter can be produced by using processes such as die-casting, extrusion and rolling; thus resulting in differing ductilities.

A study of the rivetability of magnesium alloys to aluminium alloys showed that, when used as the top material, die-cast AM50 magnesium alloy could be self-piercing riveted to extruded AA6063. When the AM50 was used as the bottom material however severe cracks occurred at the joint buttons. Because AA5754 and the wrought magnesium alloy, AZ31, have large elongations of 26 and 21%, respectively, it is possible to use them as both top and bottom materials[198].

Copper sheets have excellent ductility and self-piercing riveting has been successfully used to join aluminium to copper[199] and copper to copper[200]. The process can also join magnesium to steel and copper to titanium. Titanium alloys have a limited formability at room temperature but, when heated to above 700C, can be successfully joined[201,202].

Self-piercing riveting has also been used[203] to join sandwich materials, showing that it is possible to join 0.2mm steel plus 1.6mm polymer plus 0.2mm steel to a 2mm-thick aluminium alloy. Self-piercing riveting has also been combined with adhesives in order to form magnesium joints having a higher joint strength[204].

Friction self-piercing riveting is a new process which combines self-piercing riveting with friction stir spot welding[205]. It is capable of eliminating cracks and improving joint performance. In the friction self-piercing riveting process however, the generation of transient friction heat and its effect upon the interaction between the rivet and joined sheets can be uncertain. A 3-dimensional thermomechanically-coupled finite-element model of the process was developed, and a pre-set crack-failure method was used to model the failure of the top sheet. By analyzing the differences in physical properties of typical dissimilar materials, 3 principle problems were identified[206] in the welding and joining of dissimilar materials. These were: the presence of hard and brittle phases at faying surfaces, galvanic corrosion and distortion/stress. Their relevance to mechanical joining, welding/brazing, solid-state welding and adhesive bonding was considered. An improved friction self-piecing riveting process was proposed[207] for joining dissimilar materials by using the same equipment and merely changing the rivets and dies. The process was applied to the joining of aluminium alloy to low-ductility magnesium alloy and of aluminium alloys to high-strength steel. As noted above, the formation of weakening intermetallic compounds at an Al‖Fe interface can be a serious problem. Recent studies[208] of nanoscale shear-localization induced amorphization at such interfaces have guided the development of a modified friction-stir additive-manufacturing process. This method counteracts the detrimental effects of the intermetallic compounds by creating a nanoscale amorphous phase between a deposited aluminium alloy and a steel substrate. The formation of the nanoscale amorphous metal at the Al‖Fe interface leads to ductile fracture behaviour within the aluminium during mechanical testing.

A rotational hammer rivet technique has been developed[209] for the fabrication of magnesium/aluminium joints. The heat generated by plastic deformation of a magnesium rivet creates a metallurgical bond between the rivet head and magnesium sheet which seals off corrosive electrolyte from penetrating around the rivet head.

The wide possible variety of assemblies involving sheets having differing gauges and material properties creates problems of rivet and die selection. Seven sets of rivet and die combinations, having various die/rivet volume ratios, were used[210] to investigate the effects of rivet hardness, rivet length, die width and die pip-height, upon the riveting ability and mechanical behaviors of AA6061-T6 aluminium alloy and CR4 mild steel. The lap-shear strength was found to be linearly correlated to the undercut for a given top-sheet thickness, and an equation predicted the lap-shear strength in terms of undercut and top-sheet thickness. Plots illustrated the range of riveting ability and joint quality for each combination of rivet and die. The use of softer rivets and larger dies could improve the range of riveting ability while decreasing the joint strength. Longer rivets and smaller dies could constrict the range of riveting ability range but increase the joint strength. A die/rivet volume-ratio greater than unity was required in order to avoid issues such as bottom cracking and raised rivet head.

The friction stir scribe technique[211] is a method that produces continuous overlap joints between materials having very different melting points and high-temperature flow characteristics. The method involves the use of an off-set cutting tool, at the tip of a friction stir welding pin, to create a mechanical interlock between the material interfaces.

Magnetic-pulse welding is a means for achieving the solid-phase welding of stainless-steel plate to copper or aluminium plates. The weld quality depends upon the thickness of the plates and the distance between them[212]. The use of confocal laser-scanning microscopy, scanning electron microscopy and energy-dispersion spectroscopy to compare the microstructures of weld interfaces showed that the microscopic features of the weld interfaces vary widely. With increasing plate thickness and gap-size, the wave size and thickness of the transition zone increase, leading to improved weld quality.

Vaporizing-foil actuator welding was used[213] to join four Al|Fe combinations which consisted of the high-strength alloys: AA5052-H32, AA6111-T4, DP980 and 22MnB5. Flyer-velocities of up to 727m/s were attained by using 10kJ of input energy. During lap-shear testing, the samples failed mainly in the aluminium base and at close to the strength of the aluminium. This showed that the welds were stronger than the base metal, and that the latter was not appreciably weakened by the welding process. A particularly strong area of the weld was studied using transmission electron microscopy. It was found that the microstructure here was associated with a continuously bonded fully crystalline interface, together with nanoscale grains, mesoscopic and microscopic wavy features, and an absence of large continuous intermetallic compounds.

Advanced high-strength steels are important factors in vehicle lightweighting, and crash-resistant and fracture-toughened adhesives can improve vehicle stiffness, noise, vibration and crashworthiness. They aid weight-reduction while maintaining crashworthiness. The adhesive bonding of galvanneal coated steels has however tended to fail, with the galvanneal coating peeling away from the steel substrate and resulting in poor bond-strength. The reliability of the bonding of galvanneal coated steels with a 590MPa strength-level was studied: 590R (complex phase, high yield-strength to ultimate tensile-strength ratio), 590Y (dual-phase), 590T (transformation-induced plasticity). The adhesive bonding was evaluated by performing lap-shear tests at room temperature. The bonds were made using a crash-resistant structural adhesive, a fracture-toughened structural adhesive or a conventional structural hem-flange adhesive. Lap-shear tests were also performed on galvanneal coated interstitial-free extra deep drawing steels which were bonded with the same three adhesives in order to determine whether the lap-shear performance of galvanneal can be improved by using new-generation adhesives. X-ray diffraction has used to measure the residual stresses which exist at the interface between the zinc coating and steel substrate. It was concluded crash-resistant structural adhesives can produce good bonds between galvanneal coated advanced high-strength steels, thus reducing weight while improving crash performance.

Batteries

In the case of hybrid electric vehicles, plug-in electric vehicles, extended-range electric vehicles and purely battery-powered electric vehicles, an obvious target for lightweighting is the battery and its support structures; assuming always that the maximum energy-efficiency has already been achieved. There is then the choice of either increasing the electric-only driving-range for a given level of electrochemical-energy storage, or keeping the range constant and instead reducing the level of energy storage and the related costs of that storage. This obviously leads to a compromise having to be made between battery-cost savings, resulting from the vehicle-weight reduction, for a given range and the costs which arise from lightweighting the vehicle. There will usually be a crossover point at which the overall cost of weight-reduction exceeds the savings in battery-cost. When the energy-storage costs are relatively high, as in the recent past, it is preferable to promote lightweighting but, as those costs decrease, the more expensive lightweighting schemes can become uneconomical.

The present energy-densities of batteries mean that large and expensive batteries are required in order to meet driving-range requirements, and lightweighting obviously provides lighter vehicles which can offer achieve the same required range while needing less energy and a smaller battery. The predicted costs depend upon parameters that will generally improve; such as aluminium production, battery production and battery design. In particular, battery-cost may have decreased by up to 50% by 2030. The overall cost of the vehicle will be the final arbiter of material-replacement choices. Decreases in electric-vehicle costs will occur in time, and this will widen their use and their supportive infrastructure. At the moment, there are journeys across the USA which would be impossible to complete by electric vehicle; both because of the sparsity of recharging stations in certain regions and because, even when sufficient stations appear to exist, this is a mere illusion due to non-standardization of the charger couplings.

Given that a quarter of all energy is consumed by personal transportation needs, lightweighting is imperative. Process-based cost models for the estimation of overall vehicle costs and the effect of lightweighting schemes for electric vehicles, given alternative strategies, show that high-strength steel body structures and closures will compete with aluminium-based designs. The closures (doors, hood, hatchback) may be typically based upon a mass-produced electric vehicle having a gross mass of 1433kg, with a 35kW/207kg battery and a range of 250km. The manufacturing costs of the battery, drive train and the remainder of the vehicle then have to be estimated. The higher manufacturing costs involved in producing parts, assembling them and painting the aluminium body and closures have to be offset by the cost savings that result from the battery and motor re-sizing made possible by the lighter aluminium body. Assembly of a vehicle having the above specifications may involve 93m of laser-welding, 40m of adhesive beading and 1400 spot-welds. Vehicle sub-systems are not expected to be affected by lightweighting schemes. General overheads can be assumed to be the same for parts manufacture, assembly and painting. The cost of manufacturing each part has to be estimated separately. The combination of part-design and associated manufacturing-process choice permits a prediction to be made of the time to be allocated to the production of each part. The painting step is not greatly affected by the base material, although the initial priming step can be strongly affected by the difference in the adhesion behaviours of steel and aluminium. It can be assumed that the painting costs for aluminium are 20% higher than for high-strength steel due to alkaline degreasing. All of these considerations can in turn be converted into energy and labour costs, and ultimately determines the retail cost; based for example upon an assembly-line dedicated to the production of 250000 vehicles/year.

Keeping the range constant, and decreasing the mass of the car, permits the use of a lower-capacity and cheaper battery. In order to determine the battery mass and cost for an aluminium design, it is necessary to determine the energy required to move the lighter vehicle. This can be done by simple linear regression, using typical magnitudes for the various parameters.

The energy requirement, E, of an aluminium vehicle is assumed to scale linearly by the same proportion as the difference in the vehicle test mass, T_m,

$$E \text{ (kWh/mile)} = 51.603 + 0.1615T_m$$

The battery-energy capacity, C, can be estimated from,

$$C \text{ (kWh)} = 1.14R(\text{miles}) \times E(\text{kWh/mile})$$

where R is the range and E is the required energy. The required mass of the battery is then given by,

$$m_B \ (kg) = 0.002129C^2 + 3.1947C + 36.664$$

A recent innovation has been the construction[214] of a lithium-metal battery which is based upon a lithium/carbon-fiber woven fabric anode, a LiFePO$_4$/carbon-fiber woven fabric cathode, a glass-fiber woven fabric/PEO electrolyte and a glass-fiber woven fabric/epoxy pack. The electrolyte acted as a protective layer on the lithium anode and isolated it from air. The battery, with lithium metal as the anode, had a discharge capacity of about 147mAh/g. This is close to the theoretical LiFePO$_4$ specific capacity. The charge-discharge performance remained reliable following moderate bending, tension and compression. Due to the incorporation of the carbon-fiber and glass-fiber woven fabrics, the battery also had a tensile strength of 168.4MPa and a bending strength of 157.8MPa.

Metallurgy of Lightweighting   Materials Research Forum LLC
Materials Research Foundations **133** (2022)   https://doi.org/10.21741/9781644902134

Keywords

acidification, 73, 74

bake-hardenable, 7
Bauschinger effect, 81
body torsioning, 53
body-in-white, 19, 22, 46

castability, 40
dendritic, 29

dispersoids, 43
down-sizing, 2, 44, 87
drawability, 47

elasto-viscoplastic model, 81
electroplasticity, 75
embrittlement, 48, 75, 85
eutrophication, 73, 74
extrudability, 53, 94

ferro cast iron, 5
front crash structure, 53

Hall-Heroult electrolysis, 16
Hall-Petch effect, 77
Henrob, 95
high-pressure die-casting, 14, 30, 43, 50, 63, 68, 69
hot-extrusion, 57
hydroforming, 4, 7, 72

Kock model, 71

liftgate-assembly, 62
Lomer–Cottrell lock, 82

machinability, 56, 67
Marciniak-Kuczynski, 75
metamodel, 62
microballoon, 52
monocoque, 23

nano-composite, 43
nano-scale spinodal, 43

nugget-debonding, 95

peening, 67, 87
Portevin–LeChatelier, 84
powerplant, 44
powertrain, 8, 26, 39, 41, 44

recrystallization texture, 47
recyclability, 10, 56
rheocasting, 28
rheo-diecasting, 46
rivetability, 96, 97

self-piercing riveting, 53, 54, 96-98
simulated annealing, 45
solutionizing, 45
stiffness, 5, 8, 12, 17-19, 24, 34-36, 40, 48, 49, 52, 61, 65, 66, 71, 74, 77, 87, 90, 91, 94, 99
strain-hardening, 50, 54, 77, 80, 81, 87
superlattice structures, 70
superplastic forming, 46
super-vacuum die-casting, 14, 30, 68, 69

Taycan, 6
Taylor polycrystal, 61
thermomechanical processing, 46, 61
thixomoulded, 63
Tox, 95
TRIP, 83
TWIP, 82, 83

ultralight, 7
under-hood, 46
unfoamed precursor, 49

vacuum-assisted die-casting, 30, 68
vibro-acoustic, 23

weldability, 4, 30, 88, 94

Zener effect, 59
Zener-pinning, 59

Materials Research Forum LLC
https://doi.org/10.21741/9781644902134

About the Author

Dr. Fisher has wide knowledge and experience of the fields of engineering, metallurgy and solid-state physics, beginning with work at Rolls-Royce Aero Engines on turbine-blade research, related to the Concord supersonic passenger-aircraft project, which led to a BSc degree (1971) from the University of Wales. This was followed by theoretical and experimental work on the directional solidification of eutectic alloys having the ultimate aim of developing composite turbine blades. This work led to a doctoral degree (1978) from the Swiss Federal Institute of Technology (Lausanne). He then acted for many years as an editor of various academic journals, in particular *Defect and Diffusion Forum*. In recent years he has specialized in writing monographs which introduce readers to the most rapidly developing ideas in the fields of engineering, metallurgy and solid-state physics. He is co-author of the widely-cited student textbook, *Fundamentals of Solidification*, a new (5th fully-revised) edition of which is soon to appear. Google Scholar credits him with 8489 citations and a lifetime h-index of 14.

References

[1] Islam, E.S., Moawad, A., Kim, N., Rousseau, A., World Electric Vehicle Journal, 12[2] 2021, 84. https://doi.org/10.3390/wevj12020084

[2] Fox-Penner, P., Gorman, W., Hatch, J., Energy Policy, 122, 2018, 203-213. https://doi.org/10.1016/j.enpol.2018.07.033

[3] Li, S., Li, N., Gao, Y., Li, J., International Conference on Biomedical Engineering and Biotechnology, 2012, 1889-1892.

[4] Hao, H., Wang, S., Liu, Z., Zhao, F., Energy, 94, 2016, 755-765. https://doi.org/10.1016/j.energy.2015.11.051

[5] Lovins, A.B., Cramer, D.R., International Journal of Vehicle Design, 35[1-2] 2004, 50-85. https://doi.org/10.1504/IJVD.2004.004364

[6] Wolfram, P., Tu, Q., Heeren, N., Pauliuk, S., Hertwich, E.G., Journal of Industrial Ecology, 25[2] 2021, 494-510. https://doi.org/10.1111/jiec.13067

[7] Triebe, M.J., Zhao, F., Sutherland, J.W., Journal of Manufacturing Systems, 62, 2022, 668-680. https://doi.org/10.1016/j.jmsy.2022.02.003

[8] Toensmeier, P.A., Modern Plastics, 66[9] 1989, 44-45.

[9] Mohrbacher, H., Spöttl, M., Paegle, J., Advances in Manufacturing, 3[1] 2015, 3-18. https://doi.org/10.1007/s40436-015-0101-x

[10] Pandey, A.K., Walunj, B.S., Date, P.P., Procedia Manufacturing, 15, 2018, 915-922. https://doi.org/10.1016/j.promfg.2018.07.405

[11] Cecchel, S., Chindamo, D., Collotta, M., Cornacchia, G., Panvini, A., Tomasoni, G., Gadola, M., International Journal of Life Cycle Assessment, 23[10] 2018, 2043-2054. https://doi.org/10.1007/s11367-017-1433-5

[12] Czerwinski, F., Materials 14[21] 2021, 6631. https://doi.org/10.3390/ma14216631

[13] Wu, D., Guo, F., Field, F.R., De Kleine, R.D., Kim, H.C., Wallington, T.J., Kirchain, R.E., Environmental Science and Technology, 53[18] 2019, 10560-10570. https://doi.org/10.1021/acs.est.9b00648

[14] Bushi, L., Skszek, T., Reaburn, T., International Journal of Life Cycle Assessment, 24[2] 2019, 310-323. https://doi.org/10.1007/s11367-018-1515-z

[15] Liang, J., Zhang, J., Zhu, F., Mutschler, R., Wang, Y.W., SAE Technical Papers, 2020, April.

[16] Lin, Z.C., Cheng, C.H., International Journal of Innovative Computing, Information

and Control, 6[11] 2010, 4995-5014.

[17] Seyfried, P., Taiss, E.J.M., Calijorne, A.C., Li, F.P., Song, Q.F., Advances in Manufacturing, 3[1] 2015, 19-26. https://doi.org/10.1007/s40436-015-0103-8

[18] Kangde, S., Chaudhari, V., Guttapalli, M., Londhe, A., SAE Technical Papers, 2022.

[19] Villanueva-Rey, P., Belo, S., Quinteiro, P., Arroja, L., Dias, A.C., Journal of Cleaner Production, 204, 2018, 237-246. https://doi.org/10.1016/j.jclepro.2018.09.017

[20] Sato, F.E.K., Nakata, T., Resources, Conservation and Recycling, 164, 2021, 105118. https://doi.org/10.1016/j.resconrec.2020.105118

[21] Spreafico, C., Journal of Cleaner Production, 307, 2021, 127190. https://doi.org/10.1016/j.jclepro.2021.127190

[22] Keoleian, G.A., Sullivan, J.L., MRS Bulletin, 37[4] 2012, 365-372. https://doi.org/10.1557/mrs.2012.52

[23] Raugei, M., El Fakir, O., Wang, L., Lin, J., Morrey, D., Journal of Cleaner Production, 83, 2014, 80-86. https://doi.org/10.1016/j.jclepro.2014.07.037

[24] Jhaveri, K., Lewis, G.M., Sullivan, J.L., Keoleian, G.A., Sustainable Materials and Technologies, 15, 2018, 1-8. https://doi.org/10.1016/j.susmat.2018.01.002

[25] Kandreegula, S.K., Tikoliya, J., Nishad, H., SAE Technical Papers, March, 2017.

[26] Mahadevan, R., Lecture Notes in Mechanical Engineering, Part F9, 2017, 17. https://doi.org/10.1007/978-981-10-1771-1_5

[27] Das, S., Tonn, B.E., Peretz, J.H., International Journal of Energy Technology and Policy, 2[4] 2004, 369-391. https://doi.org/10.1504/IJETP.2004.005742

[28] Grimes, O., Bastien, C., Christensen, J., Rawlins, N., Hammond, W., Bell, P., Brown, B., Beal, J., World Electric Vehicle Journal, 6[2] 2013, 452-463. https://doi.org/10.3390/wevj6020464

[29] Kim, H.J., McMillan, C., Keoleian, G., Skerlos, S.J., IEEE International Symposium on Electronics and the Environment, 2008, 4562897.

[30] Verbrugge, M., Lee, T., Krajewski, P., Sachdev, A., Bjelkengren, C., Roth, R., Kirchain, R., Materials Science Forum, 618-619, 2009, 411-418. https://doi.org/10.4028/www.scientific.net/MSF.618-619.411

[31] Wang, Z., Zhu, F., Lü, M., Chen, S., Automotive Engineering, 37[12] 2015, 1477-1481, 1476.

[32] Bi, Z., Song, L., De Kleine, R., Mi, C.C., Keoleian, G.A., Applied Energy, 146,

2015, 11-19. https://doi.org/10.1016/j.apenergy.2015.02.031

[33] Burd, J.T.J., Moore, E.A., Ezzat, H., Kirchain, R., Roth, R., Applied Energy, 283, 2021, 116269. https://doi.org/10.1016/j.apenergy.2020.116269

[34] Kim, H.C., Wallington, T.J., Environmental Science and Technology, 47[24] 2013, 14358-14366. https://doi.org/10.1021/es402954w

[35] Kim, H.C., Wallington, T.J., Sullivan, J.L., Keoleian, G.A., Environmental Science and Technology, 49[16] 2015, 10209-10216. https://doi.org/10.1021/acs.est.5b01655

[36] Kim, H.C., Wallington, T.J., Environmental Science and Technology, 50[20] 2016, 11226-11233. https://doi.org/10.1021/acs.est.6b02059

[37] Lewis, A.M., Kelly, J.C., Keoleian, G.A., Applied Energy, 126, 2014, 13-20. https://doi.org/10.1016/j.apenergy.2014.03.023

[38] Hofer, J., Wilhelm, E., Schenler, W., Journal of Sustainable Development of Energy, Water and Environment Systems, 2[3] 2014, 284-295. https://doi.org/10.13044/j.sdewes.2014.02.0023

[39] Srinivas, G.R., Deb, A., Sanketh, R., Gupta, N.K., Procedia Engineering, 173, 2017, 623-630. https://doi.org/10.1016/j.proeng.2016.12.118

[40] Wang, T., Dong, R., Zhang, S., Qin, D., Journal of Physics: Conference Series, 1670[1] 2020, 012004. https://doi.org/10.1088/1742-6596/1670/1/012004

[41] Wang, S., Liu, X., Wu, Y., Jin, G., China Mechanical Engineering, 22[16] 2011, 2001-2006.

[42] Zhou, Q., Xia, Y., Nie, B.B., Huang, Y., Lai, X.H., China Journal of Highway and Transport, 32[7] 2019, 1-14.

[43] Liu, Y., Hu, D., Liu, X., Shan, Y., Lu, H., Automotive Engineering, 41[8] 2019, 892-895.

[44] Jiang, W., Vlahopoulos, N., Castanier, M.P., Thyagarajan, R., Mohammad, S., Case Studies in Mechanical Systems and Signal Processing, 2, 2015, 19-28. https://doi.org/10.1016/j.csmssp.2015.10.001

[45] Li, Y., Lin, J., Automotive Engineering, 36[6] 2014, 763-767, 772.

[46] Liang, J., Powers, J., Stevens, S., SAE International Journal of Materials and Manufacturing, 12[1] 2018, 19-30. https://doi.org/10.4271/05-12-01-0002

[47] Hong, Q., Liu, W., Zhou, D., Automotive Engineering, 39[2] 2017, 232-236.

[48] Praveen, K., Dharani, B., Velmurugan, R., Balasubramanian, M., Shankar, V., Akella, S., FISITA World Automotive Congress, October, 2018.

[49] Wang, D., Li, S., Automotive Engineering, 43[1] 2021, 121-128, 144.

[50] Wang, D., Cai, K., Ma, M., Zhang, S., Automotive Engineering, 40[5] 2018, 610-616, 624.

[51] He, L., Zhao, J., Gu, X., Automotive Engineering, 42[6] 2020, 832-839, 846.

[52] Zhang, H., Lü, X., Zhou, D., Xia, L., Gu, X., Automotive Engineering, 42[2] 2020, 222-227, 277.

[53] Sun, G., Zhang, H., Lu, G., Guo, J., Cui, J., Li, Q., Materials and Design, 118, 2017, 175-197. https://doi.org/10.1016/j.matdes.2016.12.073

[54] Wang, Q., Huang, Z., Automotive Engineering, 41[5] 2019, 545-549, 563.

[55] Xu, Z., Xu, X., Zhang, Z., Wan, X., Zhao, Q., Automotive Engineering, 36[3] 2014, 293-297.

[56] Guo, P., Xu, C., Liu, Z., Fang, X., Li, L., Automotive Engineering, 39[2] 2017, 138-144.

[57] Shi, Y., Zhu, P., Zhang, Y., Pan, F., Automotive Engineering, 32[9] 2010, 757-762.

[58] Doke, P., Fard, M., Jazar, R., Procedia Engineering, 49, 2012, 287-293. https://doi.org/10.1016/j.proeng.2012.10.139

[59] Olatunbosun, O.A., Gauchia, A., Boada, M.J.L., Diaz, V., Proceedings of the Institution of Mechanical Engineers D, 225[2] 2011, 167-177. https://doi.org/10.1243/09544070JAUTO1556

[60] Zheng, L.F., Wang, T., Li, G.X., Journal of Vibroengineering, 19[6] 2017, 4573-4589. https://doi.org/10.21595/jve.2017.18308

[61] Long, J., Yuan, Z., Fu, X., Zhou, S., Automotive Engineering, 37[4] 2015, 466-471.

[62] Koprivc, A., Advanced Materials and Processes, 172[6] 2014, 22-25.

[63] Karakoyun, F., Kiritsis, D., Martinsen, K., Metallurgical Research and Technology, 111[3] 2014, 137-146. https://doi.org/10.1051/metal/2014034

[64] He, X., Kim, H.C., Wallington, T.J., Zhang, S., Shen, W., De Kleine, R., Keoleian, G.A., Ma, R., Zheng, Y., Zhou, B., Wu, Y., Resources, Conservation and Recycling, 152, 2020, 104497. https://doi.org/10.1016/j.resconrec.2019.104497

[65] Hofer, J., Wilhelm, E., Schenler, W., World Electric Vehicle Journal, 5[3] 2012, 751-762. https://doi.org/10.3390/wevj5030751

[66] Hofer, J., Wilhelm, E., Schenler, W., 26th Electric Vehicle Symposium, 2, 2012, 1303-1314.

[67] Alonso, E., Lee, T.M., Bjelkengren, C., Roth, R., Kirchain, R.E., Environmental Science and Technology, 46[5] 2012, 2893-2901. https://doi.org/10.1021/es202938m

[68] Ward, J., Gohlke, D., Nealer, R., Journal of Metals, 69[6] 2017, 1065-1070. https://doi.org/10.1007/s11837-017-2330-x

[69] Raedt, H.W., Wilke, F., Ernst, C.S., VDI Berichte, 2276, 2016, 485-502. https://doi.org/10.51202/9783181022764-485

[70] Raedt, H.W., Wilke, F., Ernst, C.S., Stahl und Eisen, 135[11] 2015, 161-169.

[71] Halonen, A., Metal Casting Design and Purchasing, 20[2] 2018, 30-33.

[72] Côté, P., Bryksi, V., Stunova, B.B., Solid State Phenomena, 285, 2019, 441-445. https://doi.org/10.4028/www.scientific.net/SSP.285.441

[73] Leem, D., Huang, L., Solomon, J., Wang, H.P., Cao, J., Minerals, Metals and Materials Series, 2022, 345-354. https://doi.org/10.1007/978-3-031-06212-4_32

[74] Doty, H.W., US Patent No. 6921512.

[75] Niu, G., Wang, Y., Zhu, L., Ye, J., Mao, J., Materials Science and Technology, 38[13] 2022, 902-911. https://doi.org/10.1080/02670836.2022.2068274

[76] Xu, Z., Zhao, Z., Li, B., Jiang, Y., Jin, Y., Chinese Patent No. 103911488.

[77] Chai, Y., Chinese Patent No. 104630548.

[78] Chai, Y., Chinese Patent No. 104630558.

[79] Chai, Y., Chinese Patent No. 104630583.

[80] Li, Y., Fan, X., Xu, T., Yin, W., Chen, S., Song, Z., Zhang, Y., World Patent No. 2020052129.

[81] Richey, C.E., British Patent No. 1311808.

[82] Wilhelm, E., Hofer, J., Schenler, W., Guzzella, L., Transport, 27[3] 2012, 237-249. https://doi.org/10.3846/16484142.2012.719546

[83] Raugei, M., Morrey, D., Hutchinson, A., Winfield, P., Journal of Cleaner Production, 108, 2015, 1168-1176. https://doi.org/10.1016/j.jclepro.2015.05.100

[84] Kulkarni, K.B., SAE Technical Papers, 2021.

[85] Medikonda, S., Sridharan, S., Acharya, S., Doyle, J., ASME International

Mechanical Engineering Congress and Exposition, 2019.

[86] Zhu, L., Li, N., Childs, P.R.N., Propulsion and Power Research, 7[2] 2018, 103-119. https://doi.org/10.1016/j.jppr.2018.04.001

[87] Eiliat, H., Urbanic, R.J., Villalpando, L.F., IIE Annual Conference and Exposition, 2015, 1692-1701.

[88] Kawajiri, K., Kobayashi, M., Sakamoto, K., Journal of Cleaner Production, 253, 2020, 119805. https://doi.org/10.1016/j.jclepro.2019.119805

[89] Ke, J., Wu, Z., Shi, W., Hu, X., Automotive Engineering, 42[8] 2020, 1131-1138.

[90] Ke, J., Shi, W., Yuan, K., Zhou, G., Automotive Engineering, 41[9] 2019, 1096-1101, 1107.

[91] Yang, A., Sun, Y., Wu, X., Si, X., Si, L., Ding, R., Hu, J., Automotive Engineering, 37[10] 2015, 1221-1225.

[92] Wang, T., Na, J., Xu, Z., Yang, Z., Automotive Engineering, 35[9] 2013, 827-831.

[93] Wu, Y.H., Liu, X.M., Wang, F., 2nd International Conference on Computational Intelligence and Natural Computing, 2, 2010, 352-355.

[94] Ma, F., Wang, Z., Yang, M., Liang, H., Wu, Z., Pu, Y., Automotive Engineering, 43[5] 2021, 776-783, 790. https://doi.org/10.1002/ceat.202100590

[95] Zhu, J., Wang, S., Lin, Y., Kou, H., Yu, L., Chen, G., Automotive Engineering, 37[12] 2015 1471-1476.

[96] Mahfoud, M., Emadi, D., Advanced Materials Research, 83-86, 2010, 571-578. https://doi.org/10.4028/www.scientific.net/AMR.83-86.571

[97] Stemper, L., Tunes, M.A., Tosone, R., Uggowitzer, P.J., Pogatscher, S., Progress in Materials Science, 124, 2022, 100873. https://doi.org/10.1016/j.pmatsci.2021.100873

[98] McKenna, L.W., Wohl, M.H., Woodbrey, J.C., SAE Technical Papers, 89, 1980, 506-516.

[99] Banthia, V.K., Miller, J.M., Valisetty, R.R., Winter, E.F.M., SAE Technical Papers, 1993, 930494.

[100] Kim, H.J., McMillan, C., Keoleian, G.A., Skerlos, S.J., Journal of Industrial Ecology, 14[6] 2010, 929-946. https://doi.org/10.1111/j.1530-9290.2010.00283.x

[101] Kim, H.J., Keoleian, G.A., Skerlos, S.J., Journal of Industrial Ecology, 15[1] 2011, 64-80. https://doi.org/10.1111/j.1530-9290.2010.00288.x

[102] Tan, Y., Yang, J., Wang, S., China Mechanical Engineering, 21[23] 2010, 2887-2892. https://doi.org/10.3901/CJME.2010.01.021

[103] Jordan, S., Debruin, M., Brown, C., Gasvoda, H., SAE Technical Papers, 2020. April.

[104] Nturanabo, F., Masu, L.M., Govender, G., Materials Science Forum, 828-829, 2015, 485-491. https://doi.org/10.4028/www.scientific.net/MSF.828-829.485

[105] Liu, G., Müller, D.B., Journal of Cleaner Production, 35, 2012, 108-117. https://doi.org/10.1016/j.jclepro.2012.05.030

[106] Das, S., SAE International Journal of Materials and Manufacturing, 7[3] 2014, 588-595. https://doi.org/10.4271/2014-01-1004

[107] Czerwinski, F., Materials 14[21] 2021, 6631. https://doi.org/10.3390/ma14216631

[108] Guba, P., Gesing, A.J., Sokolowski, J., Conle, A., Sobiesiak, A., Das, S.K., Journal of Achievements in Materials and Manufacturing Engineering, 84[1] 2017, 5-22. https://doi.org/10.5604/01.3001.0010.7564

[109] Guba, P., Gesing, A., Sokolowski, J., Conle, A., Sobiesiak, A., Das, S., Kasprzak, M., Journal of Achievements in Materials and Manufacturing Engineering, 92[1-2] 2019, 5-12. https://doi.org/10.5604/01.3001.0013.3182

[110] Pan, Y.P., Zhang, Z.F., Li, B., Yang, B.C., Xu, J., Materials Science Forum, 817, 2015, 127-131. https://doi.org/10.4028/www.scientific.net/MSF.817.127

[111] Wang, Z., Sun, X., Iang, S., Lu, L., Li, J., Automotive Engineering, 37[3] 2015, 366-369.

[112] Lombardi, A., Ravindran, C., MacKay, R., Journal of Materials Engineering and Performance, 24[6] 2015, 2179-2184. https://doi.org/10.1007/s11665-015-1419-z

[113] Cecchel, S., Ferrario, D., Metallurgia Italiana, 108[6] 2016, 41-44.

[114] Cecchel, S., Ferrario, D., Metallurgia Italiana, 108[6] 2016, 29-32.

[115] Kohar, C.P., Zhumagulov, A., Brahme, A., Worswick, M.J., Mishra, R.K., Inal, K., International Journal of Impact Engineering, 95, 2016, 17-34. https://doi.org/10.1016/j.ijimpeng.2016.04.004

[116] Ma, C., Lan, F., Chen, J., Automotive Engineering, 39[4] 2017, 432-439, 456.

[117] Fang, Y., Xu, J., Gao, Z., Zhang, Z., Chinese Journal of Rare Metals, 41[11] 2017, 1180-1187.

[118] Lall, C., Proceedings of the 2017 International Conference on Powder Metallurgy

and Particulate Materials, 2017, 912-929.

[119] Samuel, E., Minerals, Metals and Materials F, 4, 2018, 301-306. https://doi.org/10.1007/978-3-319-72284-9_41

[120] Sun, W., Zhu, Y., Marceau, R., Wang, L., Zhang, Q., Gao, X., Hutchinson, C., Science, 363[6430] 2019, 972-975. https://doi.org/10.1126/science.aav7086

[121] Meagher, R.C., Hayne, M.L., DuClos, J., Davis, C.F., Lowe, T.C., Ungár, T., Arfaei, B., Minerals, Metals and Materials Series, 2019, 1507-1513. https://doi.org/10.1007/978-3-030-05864-7_190

[122] Lombardi, A., Byczynski, G., Wu, C., Zeng, X., Shankar, S., Birsan, G., Sadayappan, K., Materials Science and Technology, 2019, 1049-1056.

[123] Wang, W., Xu, C., Wang, Z., Wang, Z., Li, L., Automotive Engineering, 41[6] 2019, 607-614, 640.

[124] Ibarretxe, U., Galdos, L., Otegi, N., Ortubay, R., Argarate, U., Aranburu, A., AIP Conference Proceedings, 2113, 2019, 170021.

[125] Huo, W., Sun, T., Lei, C., Wu, H., Materials China, 39[12] 2020, 924-933.

[126] Matli, P.R., Sheng, J.G.Y., Parande, G., Manakari, V., Chua, B.W., Wong, S.C.K., Gupta, M., Crystals, 10[10] 2020, 1-11. https://doi.org/10.3390/cryst10100917

[127] Kulkarni, S., Edwards, D.J., Chapman, C., Hosseini, M.R., Owusu-Manu, D.G., Journal of Engineering, Design and Technology, 17[1] 2019, 230-249. https://doi.org/10.1108/JEDT-09-2018-0154

[128] Easton, M., Beer, A., Barnett, M., Davies, C., Dunlop, G., Durandet, Y., Blacket, S., Hilditch, T., Beggs, P., Journal of Metals, 60[11] 2008, 57-62. https://doi.org/10.1007/s11837-008-0150-8

[129] Forsmark, J.H., McCune, R.C., Giles, T., Audette, M., Snowden, J., Stalker, J., Morey, M., O'Keefe, M., Castano, C., Magnesium Technology, 2015, 333-338. https://doi.org/10.1002/9781119093428.ch62

[130] Ravichandran, N., Reddy Mungara, S., Materials Today - Proceedings, 47, 2021, 4838-4843. https://doi.org/10.1016/j.matpr.2021.06.080

[131] Mirza, F.A., Chen, D.L., Fatigue and Fracture of Engineering Materials and Structures, 37[8] 2014, 831-853. https://doi.org/10.1111/ffe.12198

[132] Chen, D., Structural Integrity, 8, 2019, 126-132. https://doi.org/10.1007/978-3-030-21894-2_25

[133] Rodriguez, A.K., Ayoub, G., Kridli, G., Zbib, H., Physics Procedia, 55, 2014, 46-52. https://doi.org/10.1016/j.phpro.2014.07.008

[134] Gao, Y., Liu, H., Wan, D., Gao, D., Automotive Engineering, 33[2] 2011, 167-171, 161.

[135] Kulkarni, S., Edwards, D.J., Parn, E.A., Chapman, C., Aigbavboa, C.O., Cornish, R., Journal of Engineering, Design and Technology, 16[6] 2018, 869-888. https://doi.org/10.1108/JEDT-03-2018-0042

[136] Weiler, J.P., Journal of Magnesium and Alloys, 7[2] 2019, 297-304. https://doi.org/10.1016/j.jma.2019.02.005

[137] Jiang, X., Hu, X., Liu, H., Ju, D., Fukushima, Y., Zhang, Z., Multidiscipline Modeling in Materials and Structures, 17[5] 2021, 882-894. https://doi.org/10.1108/MMMS-09-2020-0236

[138] Jiang, X., Lyu, R., Fukushima, Y., Otake, M., Ju, D.Y., IOP Conference Series - Materials Science and Engineering, 372[1] 2018, 012048. https://doi.org/10.1088/1757-899X/372/1/012048

[139] Tharumarajah, A., Koltun, P., Journal of Cleaner Production, 15[11-12] 2007, 1007-1013. https://doi.org/10.1016/j.jclepro.2006.05.022

[140] Brister, K.E., Horstemeyer, M.F., Whitt, C.L., Fang, H., SAE Technical Papers, 2007.

[141] Ren, L., Fan, L., Zhou, M., Guo, Y., Zhang, Y., Boehlert, C.J., Quan, G., International Journal of Lightweight Materials and Manufacture, 1[2] 2018, 81-88. https://doi.org/10.1016/j.ijlmm.2018.05.002

[142] Buchanan, C.A., Charara, M., Sullivan, J.L., Lewis, G.M., Keoleian, G.A., Transportation Research D, 62, 2018, 418-432. https://doi.org/10.1016/j.trd.2018.03.011

[143] Galos, J., Sutcliffe, M., Cebon, D., Piecyk, M., Greening, P., Transportation Research D, 41, 2015, 40-49. https://doi.org/10.1016/j.trd.2015.09.010

[144] Mistry, P.J., Johnson, M.S., Proceedings of the Institution of Mechanical Engineers F, 234[9] 2020, 958-968. https://doi.org/10.1177/0954409719877774

[145] Mistry, P.J., Johnson, M.S., Li, S., Bruni, S., Bernasconi, A., Composite Structures, 267, 2021, 113851. https://doi.org/10.1016/j.compstruct.2021.113851

[146] Mistry, P.J., Johnson, M.S., Galappaththi, U.I.K., Proceedings of the Institution of Mechanical Engineers F, 235[3] 2021, 390-402.

https://doi.org/10.1177/0954409720925685

[147] Mistry, P.J., Johnson, M.S., McRobie, C.A., Jones, I.A., Journal of Composites Science, 5[3] 2021, 77. https://doi.org/10.3390/jcs5030077

[148] Matsika, E., O'Neill, C., Grasso, M., De Iorio, A., Proceedings of the Institution of Mechanical Engineers F, 232[2] 2018, 495-513. https://doi.org/10.1177/0954409716677075

[149] Winnett, J., Hoffrichter, A., Iraklis, A., McGordon, A., Hughes, D.J., Ridler, T., Mallinson, N., Proceedings of the Institution of Civil Engineers: Transport, 170[4] 2017, 231-242. https://doi.org/10.1680/jtran.16.00038

[150] Lowrie, J., Pang, H., Ngaile, G., Journal of Manufacturing Processes, 28, 2017, 523-530. https://doi.org/10.1016/j.jmapro.2017.04.021

[151] Korter, W., Ton, W., Archive fuer Eisenhuettenwesen, 7, 1933, 365-366.

[152] Frommeyer, G., Drewes, E.J., Engl, B., Revue de Metallurgie - Cahiers D'Informations Techniques, 97[10] 2000, 1245-1253. https://doi.org/10.1051/metal:2000110

[153] Anderson, D., Chernye Metally, 8, 2016, 59-67.

[154] Adam, H., Osburg, B., AutoTechnology, 4[10] 2004, 50-52. https://doi.org/10.1007/BF03246849

[155] Min, J., Lin, J., SAE International Journal of Materials and Manufacturing, 4[1] 2011, 1147-1154. https://doi.org/10.4271/2011-01-1057

[156] De Cooman, B.C., Kwon, O., Chin, K.G., Materials Science and Technology, 28[5] 2012, 513-527. https://doi.org/10.1179/1743284711Y.0000000095

[157] Parareda, S., Casellas, D., Lara, A., Mateo, A., International Journal of Fatigue, 156, 2022, 106643. https://doi.org/10.1016/j.ijfatigue.2021.106643

[158] Hardwick, A.P., Outteridge, T., International Journal of Life Cycle Assessment, 21[11] 2016, 1616-1623. https://doi.org/10.1007/s11367-015-0967-7

[159] Xia, D., Di, S., Pan, L., Zhao, Z., Lu, J., Zhang, J., Wu, C., Automotive Engineering, 43[2] 2021, 248-252, 295.

[160] Zhou, Q., Xia, Y., Wei, X., Meng, Y., International Journal of Impact Engineering, 159, 2022, 104054. https://doi.org/10.1016/j.ijimpeng.2021.104054

[161] Zhang, P., Kohar, C.P., Brahme, A., Choi, S.H., Mishra, R.K., Inal, K., IOP Conference Series - Materials Science and Engineering, 418[1] 2018, 012002.

https://doi.org/10.1088/1757-899X/418/1/012002

[162] Choi, D.Y., Uhm, S.H., Enloe, C.M., Lee, H., Kim, G., Horvath, C., Materials Science and Technology Conference and Exhibition, 1, 2017, 454-462.

[163] Lu, Q., Zheng, J., Huang, G., Wu, Y., Ding, H., Wang, Y., Journal of Pressure Vessel Technology, 143[2] 2021, 021506. https://doi.org/10.1115/1.4049874

[164] Yakubtsov, A., Ariapour, A., Perovic, D.D., Acta Materialia, 47[4] 1999, 1271-1279. https://doi.org/10.1016/S1359-6454(98)00419-4

[165] Frommeyer, G., Brux, U., Neumann, P., ISIJ International, 43[3] 2003, 438-446. https://doi.org/10.2355/isijinternational.43.438

[166] Alain, S., Chateau, J.P., Bouaziz, O., Migot, S., Guelton, N., Materials Science and Engineering A, 387-389, 2004, 158-162. https://doi.org/10.1016/j.msea.2004.01.059

[167] Dumay, A., Chateau, J.P., Alain, S., Migot, S., Bouaziz, O., Materials Science and Enginering A, 483-484, 2008, 184-187. https://doi.org/10.1016/j.msea.2006.12.170

[168] Zheng, S., Xu, H., Feng, J., Zheng, Z., Wang, Y., Lu, L., Chinese Journal of Mechanical Engineering, 24[6] 2011, 1111-1115. https://doi.org/10.3901/CJME.2011.06.1111

[169] Zhang, Y., Wang, G., Wang, J., Advanced Materials Research, 346, 2012, 483-489. https://doi.org/10.4028/www.scientific.net/AMR.346.483

[170] Chen, R., Lou, M., Li, Y., Carlson, B.E., Journal of Manufacturing Processes, 48, 2019, 31-43. https://doi.org/10.1016/j.jmapro.2019.10.010

[171] Lu, Q., Eizadjou, M., Wang, J., Ceguerra, A., Ringer, S., Zhan, H., Wang, L., Lai, Q., Metallurgical and Materials Transactions A, 50[9] 2019, 4067-4074. https://doi.org/10.1007/s11661-019-05335-5

[172] Norgate, T.E., Rajakumar, V., Trang, S., Australasian Institute of Mining and Metallurgy Publication Series, 2, 2004, 105-112.

[173] Jones, J., Kuttolamadom, M., Mears, L., Kurfess, T., Funk, K., SAE International Journal of Materials and Manufacturing, 5[1] 2012, 247-259. https://doi.org/10.4271/2012-01-0784

[174] Kuttolamadom, M., Jones, J., Mears, L., Kurfess, T., Funk, K., SAE International Journal of Materials and Manufacturing, 5[1] 2012, 260-269. https://doi.org/10.4271/2012-01-0785

[175] Shi, L., Kang, J., Chen, X., Haselhuhn, A.S., Sigler, D.R., Carlson, B.E., Procedia Structural Integrity, 17, 2019, 355-362.
https://doi.org/10.1016/j.prostr.2019.08.047

[176] Dong, P., Welding Journal, 99[2] 2020, 39S-51S.
https://doi.org/10.29391/2020.99.004

[177] Sun, L., Transactions of the Chinese Society of Agricultural Machinery, 37[9] 2006, 20-25.

[178] Kunc, V., Erdman, D., Klett, L., International SAMPE Symposium and Exhibition Proceedings, 49, 2004, 3926-3939.

[179] Kunc, V., Erdman, D., Klett, L., International SAMPE Technical Conference, 2004, 3943-3956.

[180] Lu, Y., Zhang, K., Tran, J., Mayton, E., Kimchi, M., Zhang, W., Welding Journal, 98[9] 2019, 273S-282S.

[181] Lu, Y., Sage, D.D., Fink, C., Zhang, W., Science and Technology of Welding and Joining, 25[3] 2020, 218-227. https://doi.org/10.1080/13621718.2019.1667051

[182] Deng, L., Li, Y.B., Carlson, B.E., Sigler, D.R., Welding Journal, 97[4] 2018, 120s-132s.

[183] Shi, L., Kang, J., Sigler, D.R., Haselhuhn, A.S., Carlson, B.E., International Journal of Fatigue, 119, 2019, 185-194. https://doi.org/10.1016/j.ijfatigue.2018.08.022

[184] Chen, S., Li, G., Cui, J., Automotive Engineering, 40[7] 2018, 865-869.

[185] Watson, B., Nandwani, Y., Worswick, M.J., Cronin, D.S., International Journal of Adhesion and Adhesives, 95, 2019, 102421.
https://doi.org/10.1016/j.ijadhadh.2019.102421

[186] Larose, S., Guérin, M., Wanjara, P., Materials Science Forum, 783-786, 2014, 2839-2844. https://doi.org/10.4028/www.scientific.net/MSF.783-786.2839

[187] Ross, K., Reza-E-Rabby, M., Mcdonnell, M., Whalen, S.A., MRS Bulletin, 44[8] 2019, 613-618. https://doi.org/10.1557/mrs.2019.181

[188] Chowdhury, S.H., Chen, D.L., Bhole, S.D., Cao, X., Wanjara, P., Materials Science and Engineering A, 562, 2013, 53-60. https://doi.org/10.1016/j.msea.2012.11.039

[189] Chowdhury, S.H., Chen, D.L., Bhole, S.D., Cao, X., Wanjara, P., Materials Science and Engineering A, 556, 2012, 500-509.
https://doi.org/10.1016/j.msea.2012.07.019

[190] Haraga, K., Welding in the World, 44[4] 2000, 23-27.

[191] Abe, Y., Kato, T., Mori, K., Journal of Materials Processing Technology, 177[1-3] 2006, 417-421. https://doi.org/10.1016/j.jmatprotec.2006.04.029

[192] Chung, C.S., Kim, H.K., Fatigue and Fracture of Enginginering Materials and Structures, 39[9] 2016, 1105-1114. https://doi.org/10.1111/ffe.12419

[193] Mori, K., Kato, T., Abe, Y., Ravshanbek, Y., CIRP Annals, 55[1] 2006, 283-286. https://doi.org/10.1016/S0007-8506(07)60417-X

[194] Hoang, N.H., Porcaro, R., Langseth, M., Hanssen, A.G., International Journal of Solids and Structures, 47[3-4] 2010, 427-439. https://doi.org/10.1016/j.ijsolstr.2009.10.009

[195] Hahn, O., Horstmann, M., Materials Science Forum, 539-543, 2007,:1638-1643. https://doi.org/10.4028/www.scientific.net/MSF.539-543.1638

[196] Durandet, Y., Deam, R., Beer, A., Song, W., Blacket, S., Materials and Design, 31, 2010, S13-S16. https://doi.org/10.1016/j.matdes.2009.10.038

[197] Luo, A., Lee, T., Carter, J., SAE International Journal of Materials and Manufacturing, 4[1] 2011, 158-165. https://doi.org/10.4271/2011-01-0074

[198] Wang, J.W., Liu, Z.X., Shang, Y., Liu, A.L., Wang, M.X., Sun, R.N., Wang, P.C., Journal of Manufacturing Science and Engineering - Transactions of the ASME, 133[3] 2011, 0310091-0310099. https://doi.org/10.1115/1.4004138

[199] He, X., Zhao, L., Deng, C., Xing, B., Gub, F., Ball, A., Materials and Design, 65, 2015, 923-933. https://doi.org/10.1016/j.matdes.2014.10.002

[200] Xing, B., He, X., Wang, Y., Yang, H., Deng, C., Journal of Materials Processing Technology, 216, 2015, 28-36. https://doi.org/10.1016/j.jmatprotec.2014.08.030

[201] Zhang, X., He, X., Xing, B., Zhao, L., Lua, Y., Gu, F., Ball, A., Materials and Design, 97, 2016, 108-117. https://doi.org/10.1016/j.matdes.2016.02.075

[202] He, X., Wang, Y., Lu, Y., Zeng, K., Gu, F., Ball, A., International Journal of Advanced Manufacturing Technology, 80[9-12] 2015, 2105-2115. https://doi.org/10.1007/s00170-015-7174-3

[203] Pickin, C.G., Young, K., Tuersley, I., Materials and Design, 28[8] 2007, 2361-2365. https://doi.org/10.1016/j.matdes.2006.08.003

[204] Miyashita, Y., Teow, Y.C.J., Karasawa, T., Aoyagi, N., Otsuka, Y., Mutoh, Y/. Procedia Engineering, 10, 2011, 2532-2537. https://doi.org/10.1016/j.proeng.2011.04.417

[205] Ma, Y., Li, Y., Hu, W., Lou, M., Lin, Z., Journal of Manufacturing Science and Engineering, Transactions of the ASME, 138[6] 2016, 061007. https://doi.org/10.1115/1.4032085

[206] Li, Y., Ma, Y., Lou, M., Lei, H., Lin, Z., Journal of Mechanical Engineering, 52[24] 2016, 1-23.

[207] Ma, Y., Xian, X., Lou, M., Li, Y., Lin, Z., Procedia Engineering, 207, 2017, 950-955. https://doi.org/10.1016/j.proeng.2017.10.857

[208] Liu, F., Zhang, Y., Dong, P., Journal of Manufacturing Processes, 73, 2022, 725-735. https://doi.org/10.1016/j.jmapro.2021.11.042

[209] Wang, T., Whalen, S., Upadhyay, P., Kappagantula, K., Minerals, Metals and Materials Series, 2020, 207-212. https://doi.org/10.1007/978-3-030-36647-6_32

[210] Ma, Y., Lou, M., Li, Y., Lin, Z., Journal of Materials Processing Technology, 251, 2018, 282-294. https://doi.org/10.1016/j.jmatprotec.2017.08.020

[211] Upadhyay, P., Hovanski, Y., Carlson, B., Boettcher, E., Ruokolainen, R., Busuttil, P., Minerals, Metals and Materials Series, 2017, 147-155. https://doi.org/10.1007/978-3-319-52383-5_16

[212] Cui, J., Yuan, W., Li, G., Automotive Engineering, 39[1] 2017, 113-120.

[213] Liu, B., Vivek, A., Presley, M., Daehn, G.S., Metallurgical and Materials Transactions A: Physical Metallurgy and Materials Science, 49[3] 2018, 899-907. https://doi.org/10.1007/s11661-017-4429-7

[214] Dong, G.H., Mao, Y.Q., Wang, D.Y., Li, Y.Q., Song, S.F., Xu, C.H., Huang, P., Hu, N., Fu, S.Y., Journal of Materials Chemistry C, 10[5] 2022, 1887-1895. https://doi.org/10.1039/D1TC05601H

www.ingramcontent.com/pod-product-compliance
Lightning Source LLC
Chambersburg PA
CBHW071711210326
41597CB00017B/2437